North American Mixing Forum 2015

Core Programming Area at the 2015 AIChE Annual Meeting

Salt Lake City, Utah, USA
8-13 November 2015

ISBN: 978-1-5108-1867-5

Printed from e-media with permission by:

Curran Associates, Inc.
57 Morehouse Lane
Red Hook, NY 12571

Some format issues inherent in the e-media version may also appear in this print version.

Printed by Curran Associates, Inc. (2016)

For permission requests, please contact AIChE
at the address below.

AIChE
120 Wall Street, FL 23
New York, NY 10005-4020

Phone: (800) 242-4363
Fax: (203) 775-5177

www.aiche.org

Additional copies of this publication are available from:

Curran Associates, Inc.
57 Morehouse Lane
Red Hook, NY 12571 USA
Phone: 845-758-0400
Fax: 845-758-2634
Email: curran@proceedings.com
Web: www.proceedings.com

TABLE OF CONTENTS

420624 A New Definition of Impeller Efficiency

Abstract:

The hydraulic efficiency of impellers has been defined by Brown (MIXING XXII, 2010) as the ratio of the kinetic energy of the mean flow generated to the mechanical energy input by the impeller. This calculation assumes that the velocity of the fluid discharging through the impeller is uniform across the impeller's swept area (i.e. the mean flow divided by the swept area). In reality the velocity profile across the impeller discharge is highly non-uniform and this should be taken into account when calculating the efficiency.

When the impeller starts rotating in the fluid there is a finite time when the power input is greater than the energy dissipation rate. During this time the power input approaches the dissipation and, once they are equal, the impeller simply maintains the kinetic energy in the system which was delivered during start up.

Lattice-Boltzmann large-eddy simulations are capable of modeling the time dependency of the impeller start up allowing prediction of the power input and energy dissipation during this time. The difference between these two quantities integrated with respect to time is the kinetic energy supplied and maintained once a steady-state has been achieved.

The total energy efficiency of the impellers can be compared using a dimensionless quantity which can be formed by dividing the kinetic energy by the impeller torque.

431894 Drawdown of Liquid-Liquid Systems: Comparing Performance of Conventional Stirred Tank and Confined Impeller Stirred Tank

Monday, November 9, 2015: 8:51 AM
Salon H (Salt Lake Marriott Downtown at City Creek)

Akshay Bhalerao[1], Alexandra E. Komrakova[2], **Marcio Bezerra Machado**[1], Suzanne Kresta[3] and Fatemeh Safari[4], (1)Chemical and Materials Engineering, University of Alberta, Edmonton, AB, Canada, (2)Department of Chemical and Materials Engineering, University of Alberta, Edmonton, AB, Canada, (3)University of Alberta, Edmonton, AB, Canada, (4)University of Alberta, Edmonton, Canada

Conventional stirred tanks (ST) are widely used to produce liquid-liquid dispersions and emulsions. The flow field and energy dissipation rate throughout these tanks are highly inhomogeneous meaning that the rate of change in the drop size distribution (DSD) varies significantly with the location in the tank early in the liquid dispersion process. In the impeller region characterized by high energy dissipation rate, smaller drops are produced. At the same time, even with the Reynolds number of 20 000 the top third of the tank might not be in a fully-developed turbulent regime that leads to low levels of energy dissipate rate and, thus, very little break-up. Having homogeneous flow in the tank is crucial for industrial process, such as emulsion polymerization, where a uniform DSD is required.

A comparison study of liquid drawdown in a conventional stirred tank and a confined impeller stirred tank (CIST) was performed to demonstrate the advantages of using the latter tank for dispersion testing. The tanks are shown in the figure below. Canola oil and tap water are used as working fluids with the oil forming the dispersed phase. The dispersed phase volume fraction varies from φ=10% to 30% by volume. Sets of PBT, Intermig, A310 and Rushton turbines are used. Several parameters are determined for each experiment: the point of entrainment is defined as the impeller rotational speed when the interface between the liquids is disrupted; the fully dispersed speed gives the impeller rotational speed where complete dispersion is achieved; the time required to reach fully dispersed is also recorded; and the total mixing energy required to fully disperse the oil is the product of dissipation and mixing time. It is demonstrated that the CIST requires less energy than the stirred tank to form a complete dispersion of the immiscible liquids. For instance, for φ=10 and the PBT impeller the energy requirement for CIST is 2.5 times less than for the stirred tank.

(a) (b)

Figure. Liquid drawdown in (a) conventional stirred tank (T=0.24 m and H=T) (b) CIST (T=0.076 m and H=3T). The diameter of the Rushton turbine is equal to T/2 for both tanks. The dispersed phase volume fraction is 30%.

425205 Characterization of Mixing in the EPA Baffled Flask for Dispersion Effectiveness Testing

Abstract:

The Baffled Flask Test (BFT) has been proposed by United States Environmental Protection Agency to be adopted as the official standard protocol for testing dispersant effectiveness. The mixing energy in the baffled flask is investigated in this paper. Particle Image Velocimetry (PIV) was used to measure the water velocity in the flask placed at an orbital shaker that was rotated at seven rotation speeds: 100, 125, 150, 160, 170, 200, and 250 rpm. Two dimensional velocity field in large and small vertical cross sections of the flask for each rotation speed were obtained. The 1D energy spectra indicates a nearly isotropic flow in the BF and the existence of inertial subrange. The estimated average energy dissipation rates were in the range 7.65×10^{-3} to 4 W/kg for rotation speeds of N = 100-250 rpm, of which it is slightly larger than the one estimated by prior studies using single-point velocity measurement techniques for N = 100 and 200 rpm. The Kolmogorov scale estimated in this study for all seven rotation speeds approached the size of oil droplets observed at sea, which is 50-400 μm. The average energy dissipation rate, and Kolmogorov microscale in the flasks were correlated to the rotation speed.

INVESTIGATING THE EFFECT OF BAFFLES ON CSTR PERFORMANCE FOR LIQUID PHASE REACTION

Mohammad K.Al Mesfer and Mohd Danish*,*[1]

[*]*King Khalid University, Abha, KSA*

Abstract

In this work, the effect of baffles on the performance of CSTR was investigated under different operating conditions. The results obtained for a baffled reactor were analyzed and compared with data obtained in the absence of baffles. The operating conditions considered for investigation were temperature, feed flow rate, reactor volume, and stirrer rate. The effects of operating conditions on the reaction conversion and specific rate constant were studied for liquid phase hydrolysis of sodium hydroxide with ethyl acetate. Reaction conversions of 48.6% and a specific rate constant equal to 4.89 L/mol.sec were obtained in the presence of baffles, as compared with a reaction conversion of 48.2% and a specific rate constant of 4.78 $x10^{-3}$L/mol.sec without baffles under steady-state conditions, thus highlighting the industrial importance of baffles. Overall, baffled CSTR performed better than unbaffled CSTR under the variable conditions of temperature, feed flow rate, and stirrer rate. On the other hand, the presence of baffles had a negative effect on the process performance as the reactor volume increased. Results showed a conversion of 51.8% with baffles as compared with a conversion of 53.0% without baffles in the reactor system. These process parameters may be optimized to achieve maximum conversion and overall optimization of the chemical process for industrial-level production of ethanol and sodium acetate.

Keywords

Baffles, rate constant, CSTR, hydrolysis, saponification

---

[1]Corresponding author E-mail address: mdansh@kku.edu.sa

Introduction

Several types of reactors are used in chemical or petrochemical industries. One type of reactor used commonly in industrial processing is the continuous stirred tank reactor (CSTR). A CSTR—in the form of a single tank or, more often, as tanks in a series—is almost always operated at steady state. It is used primarily for liquid phase reactions and is assumed to be perfectly mixed. There is no time dependence or position dependence of the temperature, the concentration, or the reaction rate inside the CSTR [1].The exit stream from this reactor has the same composition as the fluid within the reactor, and this type of flow is referred as mixed flow [2].

Fat or oil hydrolysis under alkaline conditions is known as saponification [3].Hydrolysis of an ester under alkaline conditions leads to the formation of an alcohol and the salt of a carboxylic acid. Sodium acetate (CH_3COONa) and ethanol ($CH_3COOC_2H_5$) is produced by the saponification of ethyl acetate ($CH_3COOC_2H_5$) by sodium hydroxide (NaOH), although the end product is not soap. Many studies are available in the literature on the process improvement for this hydrolysis reaction.

$$CH_3COOC_2H_5 + NaOH \rightarrow CH_3COONa + C_2H_5OH$$

Saponification of ethyl acetate with sodium hydroxide is a well-known reaction and this reaction is represented by second order reaction in the literature dealing with chemical kinetics [4].Several measurement techniques [5-6] at different temperatures have been used to study the saponification. The technique depends on the conductometric measurement to determine the reaction performance was described by investigator [7].This technique, based on the conductivity measurement of product samples, prevents periodic withdrawal of samples for analysis.

Agitator vessels are generally used throughout the chemical process industry for various applications such as storing, blending and reacting materials. In addition to this, most agitated vessels use baffles, and the design of the baffled system must satisfy process objectives economically. During agitation of a low-viscosity liquid, the rotating impeller imparts tangential motion to the liquid. Without baffling, swirling motion approximates solid-body rotation in which little mixing actually occurs. The primary purpose of baffling is to convert swirling motion into a preferred flow pattern to accomplish process objectives.

On the other hand, baffling has some other effects, such as suppressing vortex formation, increasing the power input, and improving the mechanical stability [8].The presence of baffles produces axial flow. Many agitated vessels such as CSTR use standard baffling, which consists of four flat vertical plates.

Standard baffling typically provides near-optimal performance and because of the symmetric placement of the baffles around the vessel periphery, standard baffling provides a high degree of mechanical agitation. At equal power input, the turbulent blend time of an unbaffled vessel is substantially greater than that of a baffled system.

Flat-plate baffles are the norm because of their ease of manufacture and installation and the associated cost-effectiveness. Efficiency of a reactive system such as CSTR largely depends on the extent of the mixing of reactants. If the reactants are premixed, the reaction can start from the entry of the reactor. Under non-premixed conditions, it is necessary for the reactants to come into contact for the reaction to take place [9]. In most cases, continuous-type reactors use mechanical agitation to achieve adequate mixing.

By using a combination of agitators and baffles, efficient mixing can occur regardless of production capacity and viscosity of the fluid. Mechanical agitator and baffle arrangements provide better CSTR performance, resulting in more conversion of reactant to product. Baffle effects on the performance of a circulating fluidized bed reactor were also investigated [10].Except at the lowest gas velocity, the riser with baffles yields a higher solids holdup and ozone conversion in the gas phase than without baffles.

Materials and Methods

Chemicals

AR-grade chemicals were used to conduct the experiments. Ethyl acetate (with 99.5% purity supplied by Sigma Aldrich Company) and sodium hydroxide (98.0%-100% from Panreac Quimics Company) were used to carry out hydrolysis reaction. High-purity distilled water was used to prepare stock solutions of NaOH (0.1M) and $CH_3COOC_2H_5$ (0.1M).

Experimental Setup

A CSTR with an internal diameter ~0.153 meter and maximum volume of 2.0 L was used to conduct the experiments. A stainless steel coil inside the reactor is used to heat the reactants. A turbine agitator in combination with baffles provides efficient mixing and heat transfer. Glands in the reactor lid house the sensors of temperature and conductivity with service unit.

Hydrolysis of $CH_3COOC_2H_5$ with NaOH is monitored by a conductivity probe because the conductivity of a solution varies with conversion. Hot water circulation through the internal coil is used to control the temperature in the reactor and is monitored by PID (Proportional integral derivative) controller. Stirrer rate (agitation) is regulated using an agitator speed control mounted on the service unit. Feed flow rates of the reactants are controlled by two peristaltic pumps.

Experimental Procedure

Solutions of sodium hydroxide (0.1 M) and ethyl acetate (0.1 M) were prepared using high-purity distilled water to carry out experimental work at different operating conditions. Data were collected for CSTR with and without baffles. As the reactants were pumped into the reactor using a peristaltic pump, the level increased until it finally overflowed the standpipe.

After attaining the desired operating conditions in the reactor, real-time conductivity was recorded. The extent of the reaction and the specific rate constants were calculated with and without baffles using conductivity.

Results and Discussion

Steady-State Condition

Reactant flow rates of 60 ml/min NaOH and 60 ml/min $CH_3COOC_2H_5$ were adjusted using peristaltic pumps. Hydrolysis reaction was carried out at a constant temperature of 30°C, with a stirrer rate of 130 rpm and at a concentration of 0.1 M for both the reactants. Real-time conductivity was recorded at 5-minute intervals until steady-state condition was achieved under presence of baffles and without baffles. The steady-state condition was reached after a period of 35 minutes from the start of the experiment, as shown in Figure 1.

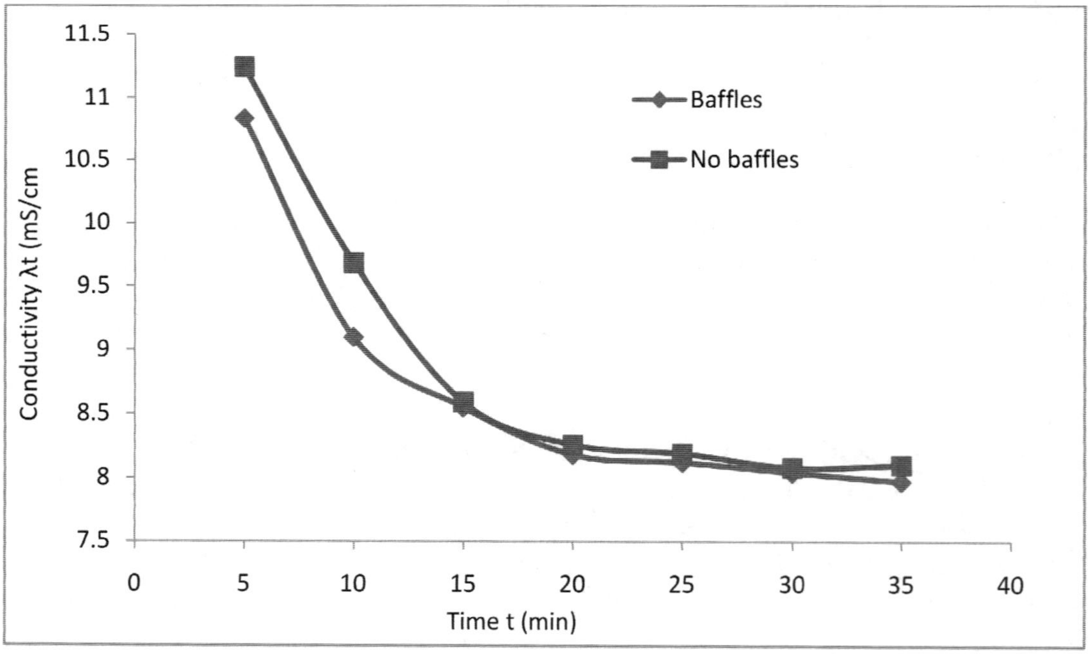

Figure 1 Conductivity versus time curves

Only sodium hydroxide and sodium acetate contribute conductance to the reaction solution. Reaction mixture conductivity decreased with time and attained a steady state after

35 minutes. Concentration of NaOH decreased as the reaction proceeded, leading to a decrease in conductivity values. A reaction conversion of 48.6% and a specific rate constant of 4.89×10^{-3}. L/mol.sec were obtained with baffles as compared with a conversion of 48.2% and a specific rate constant of 4.78×10^{-3} L/mol.sec without baffles under steady-state conditions, thus validating the industrial significance of baffles.

Temperature

This study was conducted at an agitator speed of 130 rpm and a constant flow rate of 60 ml/min for both the reactants. Reactor volume was adjusted to 1.5 Liter. Both the conversion and the specific rate constant increased with rising temperature, with and without baffles. Moreover, baffled CSTR contributed to higher conversion, as shown in Figure 2.

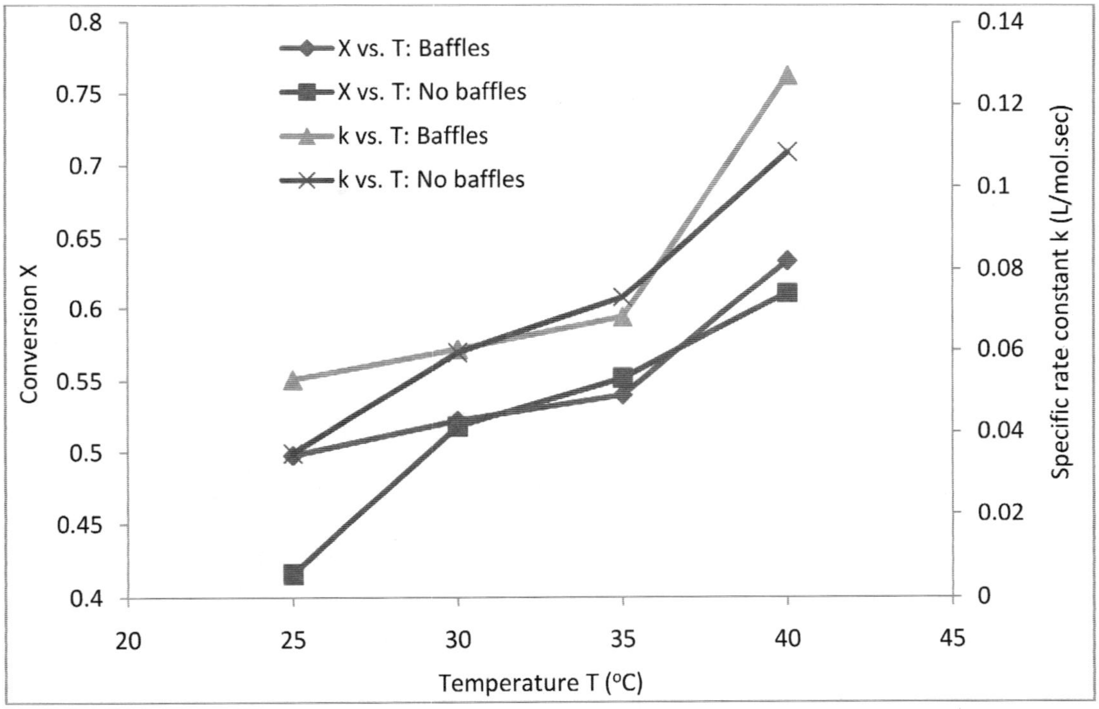

Figure 2 Conversion and Specific rate constant versus temperature curves

A higher reaction conversion of 63.4% with baffles was recorded as compared with a 61.1% in the absence of baffles under constant operating conditions of reactant flow rate, concentration, reactor volume, and stirrer rate. Specific rate constant increased with higher temperature of the reaction mixture; moreover, a higher value of k was obtained for the baffled CSTR as compared with the unbaffled CSTR at the same temperature.

Stirrer Rate

Flow rates of both reactants were adjusted to 60 ml/min. Volume of reactor and temperature were 1.5 Liter and 30°C, respectively. Conversion and specific rate constant increased with stirrer rate, as shown in Figure 3. A steady-state conversion of 55.2% was obtained for the baffled reactor as compared with a lower value of 47.5% with no baffles at an agitation rate of 190 rpm. A similar trend was noticed for k; a higher value of 4.89×10^{-3} L/min.sec was obtained in the presence of baffles in comparison with a value of 4.79×10^{-3} L/mol.sec without baffles. Agitator in concurrence with baffles is necessary for maintaining the uniform temperature throughout the reactor and hence the conversion of the reaction.

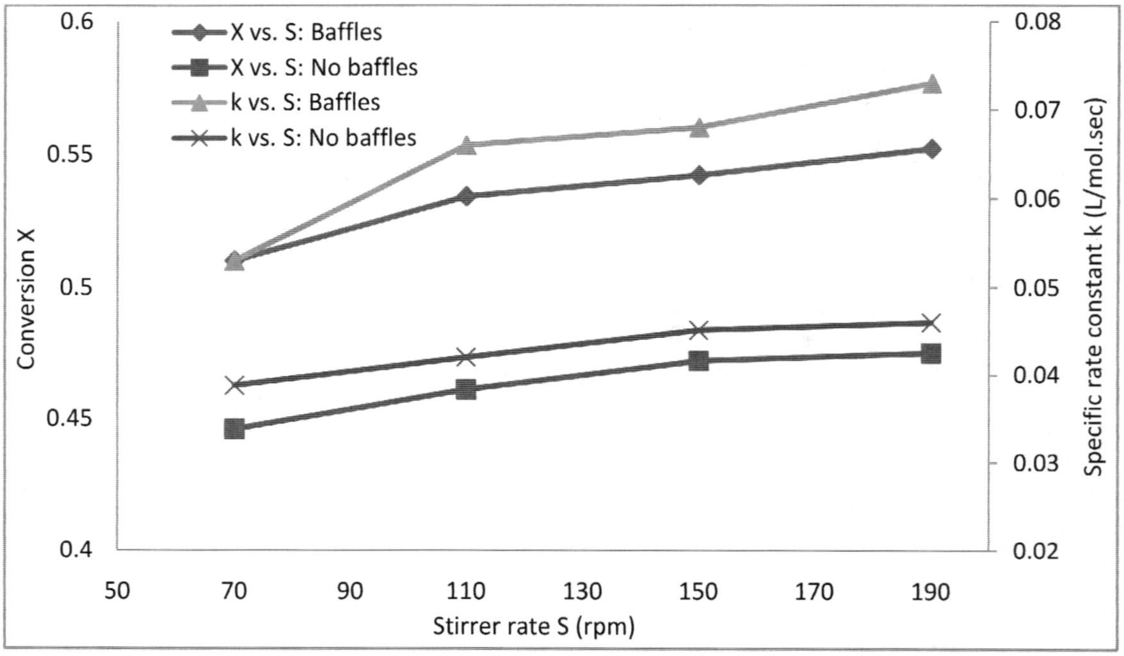

Figure 3 Conversion and specific rate constant versus stirrer rate

Reactant Flow Rate

The effect of reactant flow rates on conversion and rate constant was investigated, as shown in Figure 4. Overall, the baffled CSTR performed better as compared with the unbaffled system.

Steady-state conversion and specific rate constant of 52.8% and 5.26 L/mol.sec, respectively, were achieved in the presence of baffles as compared with lower values of 52% and 4.27 L/mol.sec, respectively, with no baffles at a constant reactant flow rate of 50 Liter/sec. It was also noted that increased reactant flow rates lead to decreased conversion values. This is due to the fact that residence time decreases with increased feed flow rates.

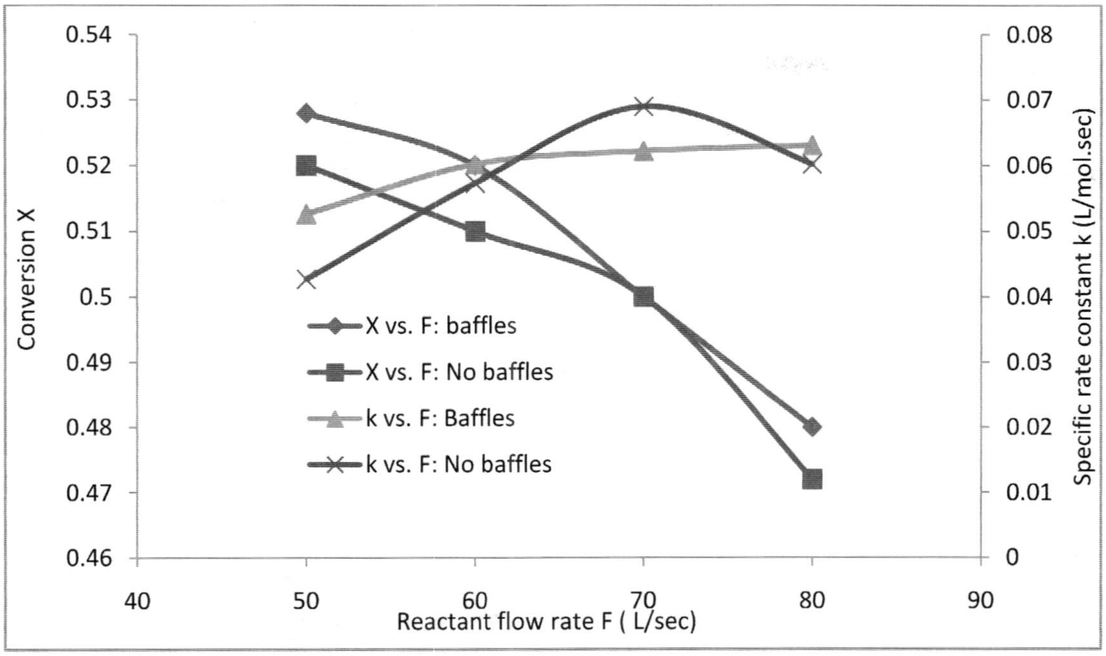

Figure 4 Conversion and specific rate constant versus flow rate curves

Reactor Volume

Data were collected to analyze the effect of reactor volume on reaction conversion and specific rate constant and results are presented in figure 5.

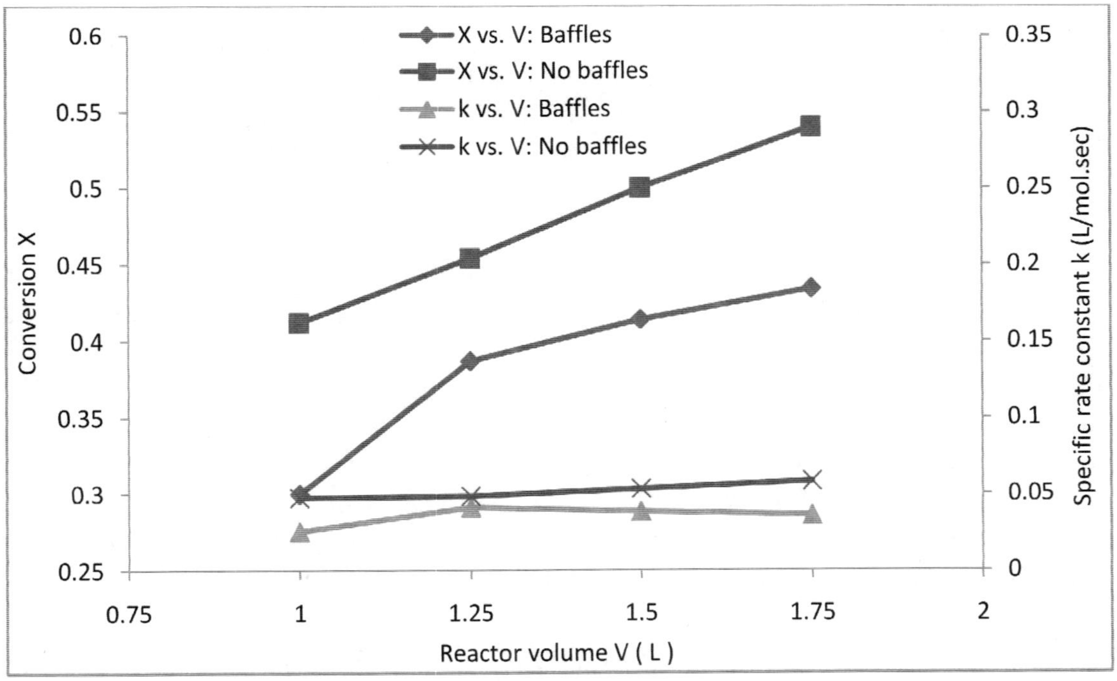

Figure 5 Conversion and rate constant versus reactor volume curves

Feed flow rates of 60 ml/min were adjusted for both reactants and a reaction temperature of 30°C was maintained at an agitation rate of 130 rpm. The presence of baffles contributed negatively to the process performance as the volume of the reactor increased. Indeed, with increased reactor volume, a conversion of 43.4% was obtained with baffles as compared with a conversion of 54.0% without baffles in the reactor system. Similar results were obtained for k; it was observed that a higher value of k ~ 3.85×10^{-3} L/mol.sec was achieved with baffles as compared with k ~ 5.33×10^{-3} L/mol.sec without baffles.

Conclusions

In this study, saponification of ethyl acetate (~0.1 M) with sodium hydroxide (~0.1M) was investigated, and the effect of baffles on the performance of CSTR was analyzed. Progression of the reaction was monitored by real-time conductivity data under different operating conditions. The results obtained for baffled reactor were analyzed and compared with data obtained in the absence of baffles.

- A decline in the value of real-time conductivity with time indicates the progress of reaction.
- Reaction conversions of 48.6% and a specific rate constant equal to 4.89 L/mol.sec were obtained in the presence of baffles in the reactor system as compared with a conversion of 48.2% and a specific rate constant of 4.78×10^{-3} L/mol.sec without baffles under steady-state condition, indicating the industrial importance of baffles.
- The conversion and specific rate constant increased with increased temperature both with and without baffles; moreover, baffled CSTR contributed to higher conversion. A reaction conversion of 63.4% with baffles was observed as compared with a lower value of 61.1% under constant operating conditions of reactant flow rate, concentration, reactor volume, and stirrer rate.
- The conversion and specific rate constant increased as a function of stirrer rate; a conversion of 54.2% and a specific rate constant of 6.80×10^{-3} L/mol.sec were obtained for a baffled reactor as compared with a lower conversion of 47.2% with no baffle at stirrer rate of 140 rpm.
- High conversion and specific rate constant were calculated for a baffled reactor as compared with an unbaffled reactor as a function of reactant flow rate. Results also showed that an increase in reactant flow rate lead to decreased conversion. This occurs because residence time decreases with increased reactant flow rate.
- On the other hand, the presence of baffles also contributed negatively to the process performance as the volume of reactor increased. With increased volume, a conversion of 51.8% with baffles was obtained as compared with a conversion of 53.0% without baffles in the reactor system.

- It is concluded that these process parameters may be optimized to achieve maximum conversion and, hence, overall optimization of the chemical process for industrial-level production of ethanol and sodium acetate.

Acknowledgement

The authors are thankful for the valuable support of the chemical engineering laboratory staff, King Khalid University, Abha, Kingdom of Saudi Arabia.

References

1. Fogler, H.S. (2008), "Elements of Chemical Reaction Engineering," Printice Hall, IV Edition, NJ, USA, pp.11-12.

2. Levenspiel, O. (1999), "Chemical Reaction Engineering," John Wiley and Sons, Third edition, NJ, USA, pp.90-91.

3. Bursali, N. Ertunc, S. and Akay, B. (2006), "Process Improvement Approach to Saponification Reaction using Statistical Experimental Design," Chemical Engineering and Process, 45, pp. 980-989.

4. Kapoor, K.L. (2004), "A Textbook of Physical Chemistry," McMilan, ND, India, Vol.5, pp. 116.

5. Daniels, F. Mattews, J. and Williams, J. (1941), "Experimental Physical Chemistry," McGraw-Hill, New York, pp. 167-169.

6. Schneider, M.A. and Stoessel, F. (2005), "Chem Eng J," 115, pp. 73-83.

7. Walker, J. (1996), "Proc. R Soc A," London, 78, pp. 157.

8. Kevin, J.M. Reeder, M.F. and Fasano, J.B. (2002), "Optimize Mixing by Using Proper Baffles," CEP magazine, AIChE Publication, pp. 42-47.

9. Rajavathasavai, D. Khapre, A. and Munshi, B. (2014), "Study of Mixing Behavior of CSTR Using CFD," Brazilian Journal of Chemical Engineering, Vol.31, No.01, pp.119-129.

10. Jiang, P. Bi, H. Jean, R.H. and Fan, L.S.(1991), "Baffle Effects on Performance of Catalytic Circulating Fluidized Bed Reactor," AIChE Journal, September, Vol.37, No.9, pp.1392-1400.

CFD ANALYSIS IN DESIGING OF MIXING MECHANISM FOR A MECHANICAL FLOTATION MACHINE

Jiliang Xia[1], Antti Rinne[1], Jarmo Lohilahti[1], Raghav Dube[2], Toni Mattsson[1]
[1] *Outotec, Beneficiation, Finland*
[2] *Outotec, USA*

ABSTRACT

This paper describes a CFD and experimental study in designing TankCell® mechanical flotation machines at the Outotec large capacity cells. A large flotation machine size ranges from 100 m^3 to 630 m3 (which is the largest flotation machine in the world). Outotec FloatForce® mixing mechanism was examined in detail with flotation performance as a function of rotor speed. Results are presented in the form of velocity plot, turbulent quantities, flow characteristics such as power draw, power number, pumping number, specific energy consumption, and others. Mixing profile measurements performed inside large scale flotation machines are discussed in accordance with CFD results. Both CFD and industrial experiments have shown that the Outotec flotation machine achieves good mixing and high recovery efficiency. It is shown that specific energy consumption becomes smaller with increasing the cell size. The increase in rotor speed enhances mixing and increases power draw, and an optimum rotor speed for specific flotation machine size is obtained in this study. The CFD reveals a spectrum of fluid flow, mixing, and particle-bubble behavior in the whole flotation machine and provides deep insight into the flotation mechanisms. Results obtained from CFD have been validated against experimental measurements from industrial-scale flotation machine.

INTRODUCTION

Computational Fluid Dynamics (CFD) has become an inevitable tool in engineering design and operation. CFD can solve most problems involved in fluid flow and heat transfer occurring in metallurgical processes. Through CFD predictions, one can better understand how a process works, improve the systems, and create new products. Outotec has being long used CFD techniques in its process and product design, improvement, optimization, and overshooting problems occurring in the existing operational systems.

Extensive experimental and CFD studies have been carried outin order to understand flow behavior in a flotation cell. Tatterson (1991) carried out a thorough discussion on liquid mixing and gas dispersion in agitated tanks and well addressed early research on mixing flows from various agitation mechanisms. Grau and Heiskanen (2003) and Grau (2006) conducted a comprehensive experimental investigation of the frother concentration effect on bubbles in an Outotec TankCell® and studied the effect of physical and chemical variables on bubble generation and coalescence. Azgomi et al. (2007) explored a possible way to characterize frothers using gas hold-up. Liu and Xu (2007) studied the role of reagents in tuning colloidal forces for sophalerite-silica separation. Laskowski (2007) carried out a thermodynamic analysis of the role of bubbles in transporting hydrophobic particles to froth phase in flotation.

CFD techniques have been intensively applied in flotation research and design. For example, Koh and Schwarz (2005) predicted the bubble-particle behavior using population balance model. Tiitinen (2003) and Zhu et al. (2002) simulated the flow in Outotec flotation cells. Xia, Rinne and Grönstrand (2009) simulated flows in the Outotec TankCell®-160 and clarified the capability of three turbulence models in predicting flotation mechanisms. They (2010) also examined the effect of the bottom clearance and the rotor speed on flotation behavior. An overview of the literature focused on using CFD for modeling froth flotation process was conducted by Sarha, Naser and Brooks (2014). Flows in a 600 m^3 flotation cell were simulated by Chen et al.

Outotec's flotation technology offers superior flotation performance in aeration, mixing, froth manipulation, and recovery. To better understand the flow performance in the cell leads to improved design of the flotation circuit, Outotec has conducted intensive experimental and CFD studies to develop and clarify the mixing behavior of Outotec's MultiMix® and FloatForce® mechanisms. A throughout investigation has been conducted for the effect of flotation machine size, rotor diameter, rotor bottom clearance, rotor speed, slurry composition, flotation machine shape, and others on the flow and mixing performance. The present paper will mainly outline some of the CFD results, with comparison to the industrial experimental data conducted on-site.

PYSICAL MODEL AND SOLUTION METHODS

In order to obtain a throughout understanding of the flow and mixing performance of Outotec's flotation MultiMix® and FloatForce® mechanisms, a series of CFD modeling has been conducted for:

1) various tankcell sizes such as TC-e1, TC-e12, TC-e30, TC-e38, TC-e60, TC-e70, TC-e100, TC-e160, TC-e300, TC-e500, and TC-e630. It is noted that TC-e630 is the largest flotation machine so far in the world.
2) various rotor bottom clearances such as 10%, 20%, and 30%.
3) difference flotation mechanisms.
4) different flotation machine shape.
5) different slurry compositions and solid particle size distributions from industrial processes.

6) different rotor speeds and sizes.

Typical physical model considered is schematically illustrated in Figure 1. The flotation machine consists of a rotor, a stator, a shaft, top cone, baffles, vortex break, and so on.

A commercial code ANSYS Fluent (2015) is used to solve the conservation equations, along with the boundary conditions. Turbulence is modeled using the standard k-ε model, and Reynolds Stress model for some cases. The whole flotation cell (360o) is taken into account in computational domain to capture possible unsymmetrical flow caused by the rotor/stator structure. Typical refined non-uniform mesh with about 1.5 - 3.5 M elements is utilized, which is fine enough to obtain mesh-independent solution for the conditions considered. The rotational effect of rotor is considered using the moving-mesh technique. For some cases, transient computation is conducted considering the unstable feature of flows in the cell. For other cases steady-state modeling is carried out. Second-order algorithm is used to discretize all derivatives. The converged solution is assumed when the scaled residuals of all variables are smaller than 10^{-4}.

For slurry flow, both the Mixture model and/or Eulerian-Eulerian model are used to simulate the liquid-particle two phase flows. One or two solid particle materials are considered. Two or three particle sizes are assumed, for example, 50, 120 and 220 μm. The drag, lift, and pressure forces are taken into account for the liquid-particle interactions. For the sake of brevity, the conservation equations and details of boundary conditions will not be described here.

Extensive industrial-scale and laboratory-scale experimental measurements are also carried out at Outotec for the mixing and flotation behavior (with and without air injection) at different flotation machine sizes from TC-e1 to TC-e500 and different operational conditions. For example, a series of measurements are performed for different rotor speeds and rotor bottom clearance at Inmet Mining's Pyhäsalmi concentration plant in Finland, in a TC-e60 with FloatForce® mechanism. In the measurements, air feed is fixed at Jg=1.0 cm/s. A total of 32 sampling rounds are taken with each setting. Power is changed randomly. Time between sampling varies randomly between 20 minutes and several days.

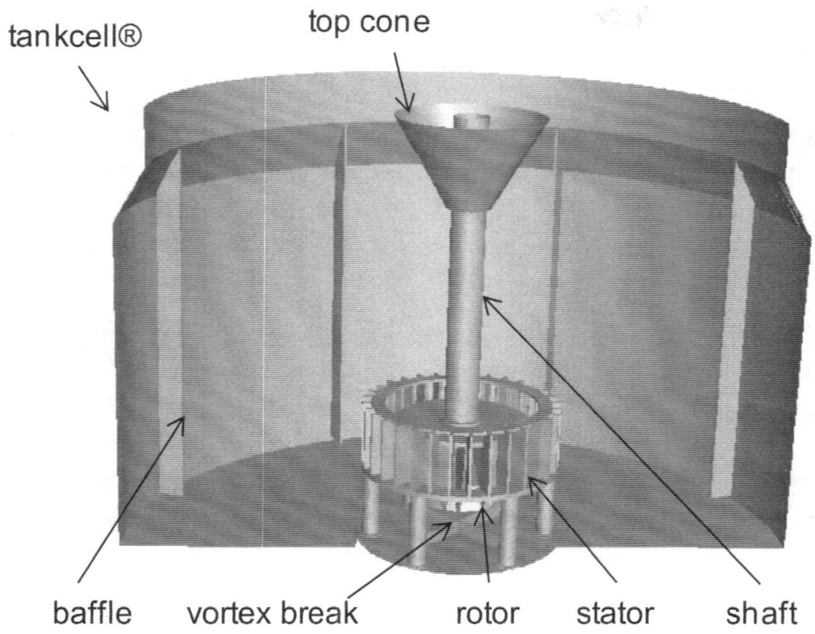

Figure 1 A schematic diagram of a typical flotation machine

RESULTS AND DISCUSSIONS

Outotec has conducted systematic CFD and experimental investigation to develop a flotation machine which can achieve excellent mixing and recovery. The present paper outlines some of Outotec's studies on flotation technology.

1) Flow and mixing behavior in flotation machines

Figure 2 shows velocity vector plots across the cell for four flotation machine sizes TC-e100, TC-e300, TC-e500, and TC-e630 at similar rotor tip speed at y = 0. It is seen that similar two-vortex flow pattern across the slurry slot is formed in different sizes. The rotating rotor pumps the slurry from the bottom into the slurry and blade slots and a jet-like flow from the slurry slot is formed, where the slurry moves radially towards the cell wall. Then the slurry stream splits into two parts. One flows downwards along the wall, back to the central region, and upwards into the rotor, forming a low vortex recirculation zone. Above it the slurry flows upwards along the wall and moves downwards from the central region to mix with the slurry stream from low recirculation zone, forming a large recirculation zone in the upper region. It is revealed that this two recirculation flow pattern across the slurry slot will not hold across the rotor blade slot, where there does not appear a jet-like flow as shown in Figure 3 which shows the velocity vectors around the rotor region. The rotor creates strong mixing and agitation. The CFD modeling clearly shows that there appears a toriodal recirculation zone within each rotor blade slot. In this vortex zone the maximum velocity can be higher than the rotor tip speed.

Figure 4 shows velocity vector plot at z=1.65 m below the rotor top cover for TC-e300, it is seen that there appear two vortexes near the side walls across each of the rotor blade slot. It is also noted that the stator blades do subdue the swirling flow induced by the rotating rotor and guide the slurry flowing nearly radially out of the machine bulk region.

The mixing behavior in a TC-e160 with the inlet, underflow, and overflow taken into account is modeled. The slurry consists of two solid materials. For each solid material two particle sizes are assumed, Figure 5 shows contour of volume fractions of hematite and silica particles with a diameter of d_p = 44 µm and d_p = 200 µm. Complex multiple vortex flow pattern appears for the slurry with solid particles taken into account, especially in the upper region of the flotation machine. It is revealed that the Outotec flotation mechanism can introduce excellent mixing. There is observed some effect of the inlet, underflow, and overflow on the location of vortex center for the feed inlet and underflow sides of the flotation cell.

The pathlines colored by the velocity magnitude is shown in Figure 6. It is clearly seen that two main vortexes surrounding the whole 360o of the flotation machine domain are formed. This figure also shows an important fact that all the feed from the inlet are sucked into the rotor region before moving to the underflow and no particles flow directly into the underflow pipes from the feed inlet. This explains why Outotec flotation machines can achieve good mixing and flotation performance.

Contour of shear stress on the rotor and stator surfaces is shown in Figure 7. The location with high shear stress is clearly demonstrated. The higher stress indicates more serious erosion. It is noted that the serous wear location revealed by the CFD prediction is in a good agreement with industrial observation.

2) Effect of rotor speed and rotor bottom clearance

Intensive CFD studies have been carried out for different operational conditions. Figure 8 shows the change of the power draw with the rotor speed for TC-e60 at a 30% rotor bottom clearance. It is seen that the power draw increases drastically with increasing the rotor speed. A higher rotor speed also causes more wear and shorter rotor life. Therefore, in practical applications it is desirable to apply an optimized rotor speed to make a compromise between the mixing behavior and erosion. Outotec has found the optimized rotor speed for each size of flotation machines through vast CFD simulations and experimental measurements.

Change of power number with the rotor speed is plotted in Figure 9 for three rotor bottom clearance. It is seen that the power number remains nearly unchanged when the rotor speed varies. The increase in the bottom clearance increases slightly the power number. In the CFD modeling are calculated the rotor pumping mass flow rate, the flow number, the pumping efficiency and effectiveness, and the specific energy consumption, respectively. It is found that

1) with increasing the rotor speed, the pumping mass flow rate increases, the ratio of the mass flow rate to power draw decreases, the mass flow rate per energy consumption decreases, and the mass flow rate against power number increases.
2) the mass flow rate through the rotor increases with increasing flotation machine size, and the rate per specific energy consumption also increases.
3) the Outotec flotation machine with Floatforce® mechanism gives the pumping number N_q~0.9, which is higher than that from other types of flotation mechanisms.
4) the pumping capacity generally slightly decreases with increasing the rotor size,
5) Similar trend to the pumping capacity is also revealed for the pumping effectiveness and the pumping efficiency.
6) The baffles appear to improve the performance of TC-e500 and TC-e630.
7) The specific energy consumption decreases with the flotation machine sizes and increases with increasing the rotor speed. The Outotec flotation mechanism achieves very low specific energy consumption, for example, it is 0.33 kW/m^3 at N=55 rpm for TC-e630.

3) Comparison of CFD solutions with experimental data

This section describes some examples of comparison of CFD predictions with the industrial-scale and laboratory-scale experimental measurements. It should be pointed out that the CFD simulations for all cases were carried out independently of experimental measurements. There has not been any adjustment of CFD models to fit to experimental data. This firmly suggests that the mathematical models used can reasonably predict flows in flotation machines.

Both CFD prediction and experimental observation reveal that there exists a toroidal recirculation zone behind each rotor blade. Figure 10 shows a comparison of the CFD solution with the experimental observation around the rotor blades. Figure 11 shows comparison of the CFD prediction with measured flow behavior around the stator. It is clearly seen that the fluid flows outwards in radial direction through the stator blade volume. The CFD modeling has reasonably predicted the flow phenomena observed in experimental observations.

To validate the numerical predictions, the predicted power draws are compared with the measured data. Comparison of power draws is made for three rotor speeds and three bottom clearances. It is shown that very good agreement is achieved for all the cases. Figure 8 shows the change of predicted power draw with the rotor speed at 30% bottom clearances, along with both measured shaft power consumption and measured electrical power consumption. The deviation between CFD prediction and experimental measurement is within –3.2% to 2.8% for the rotor bottom clearance of 30% at all rotor speeds. It is also seen that for all the bottom clearances, the measured electrical power is somewhat higher than predicted values, which is reasonable because of electric efficiency and mechanical losses.

Figure 12 shows the profile of the specific energy consumption as a function of the flotation machine size. Both CFD predictions and experimental measurements are conducted in a broad range of operational conditions such as different rotor speed, bottom clearance, blade number, and others. It is noted that the operational conditions may not be correspondingly the same for both the CFD modeling and experiments. It is seen that the specific energy consumption drastically decreases with increasing the flotation size. When the flotation machine reaches to a certain size, the slope becomes smaller with an increase in the machine size. As described in above sections, the specific energy consumption increases with increasing the rotor speed for a certain size of flotation machine. A wide range of the rotor speed is examined for a flotation machine size in order to obtain an optimized rotor speed. Some high specific energy consumption is from an assumed high rotor speed. The correlated trendlines for the both the CFD predictions and experimental data are nearly consistent. This broad comparison indicates that the CFD modeling does reveal the real flow solution and flotation behavior of a flotation machine and is a useful tool in understanding, diagnosing, designing, optimizing, and developing flotation technology.

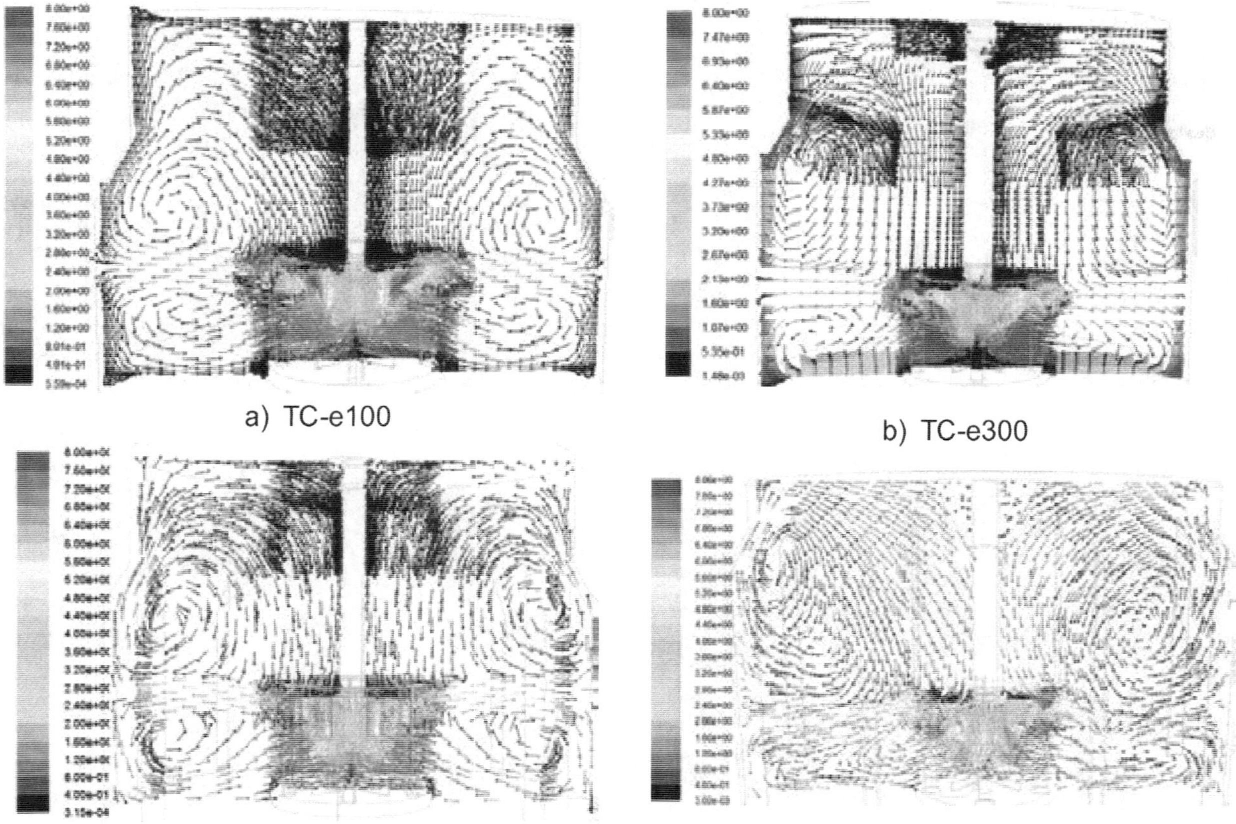

a) TC-e100

b) TC-e300

c) TC-e500

d) TC-e630

Figure 2 Velocity vector plots for four different flotation machines

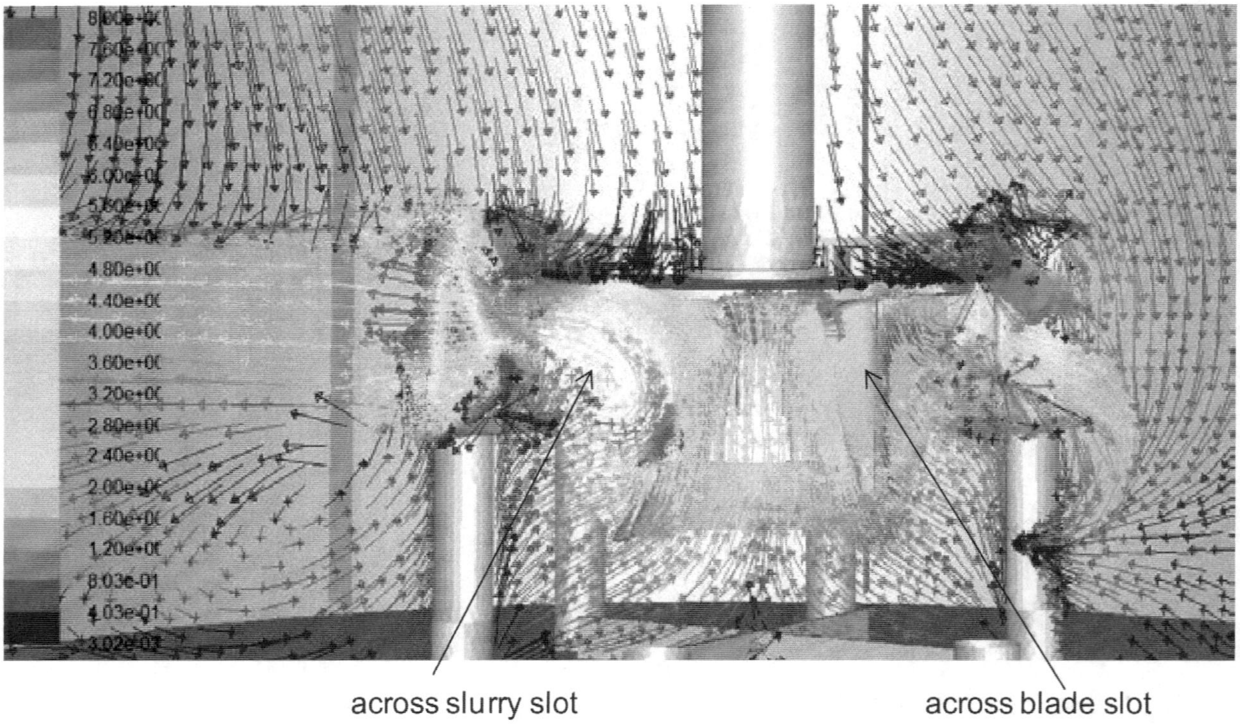

Figure 3 Flow pattern around the rotor region for TC-e630

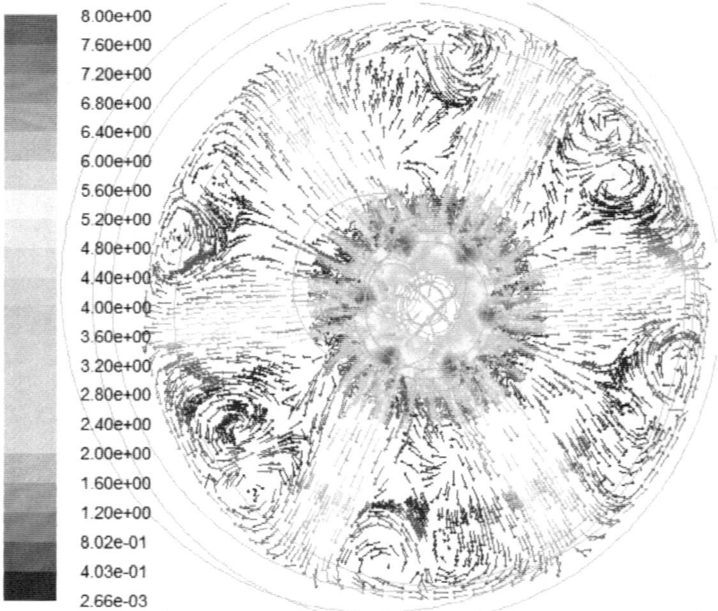

Figure 4 Flow pattern just below the rotor top cover at z=1.65 m for TC-e300

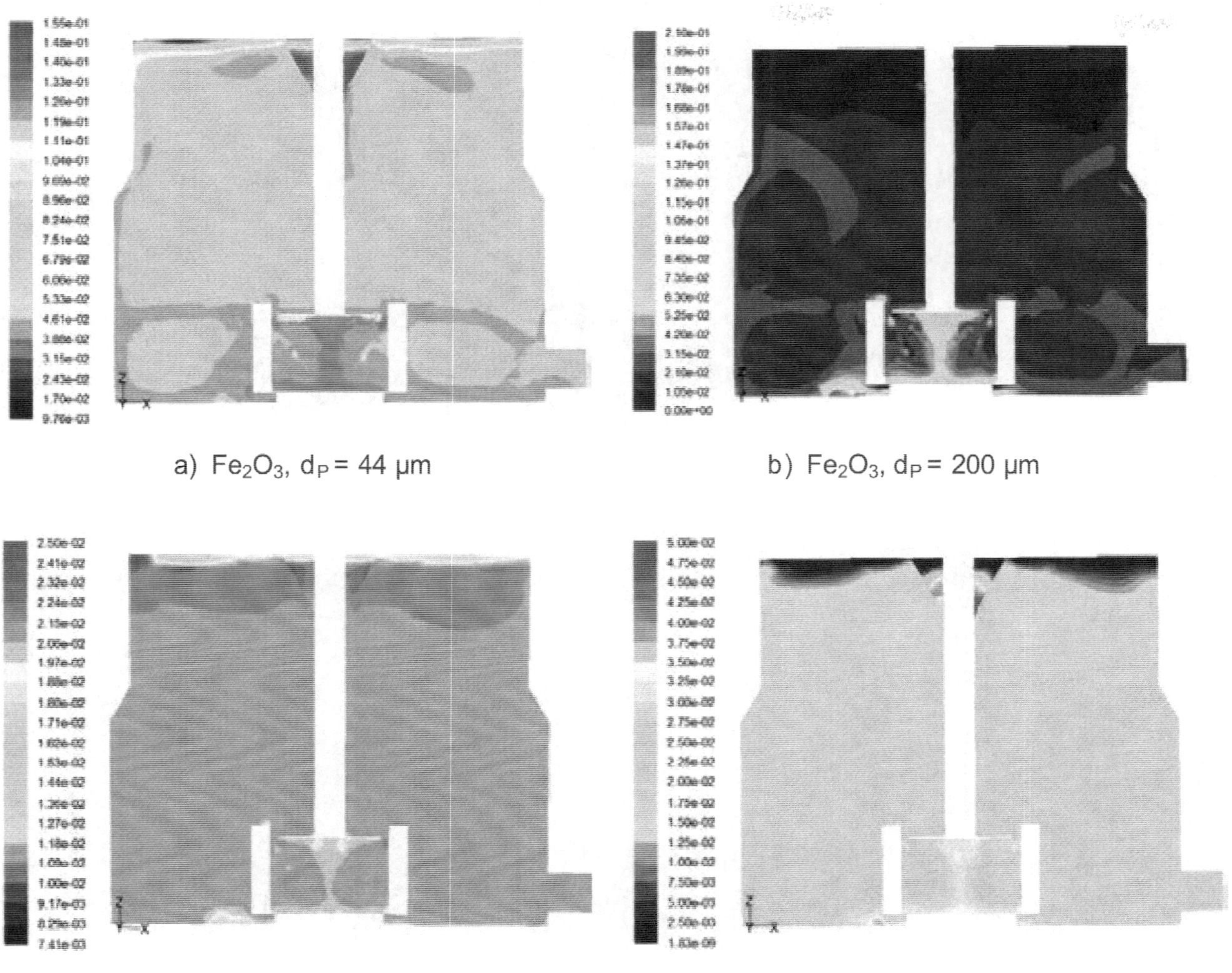

a) Fe_2O_3, d_P = 44 μm

b) Fe_2O_3, d_P = 200 μm

c) SiO_2, d_P = 44 μm

d) SiO_2, d_P = 200 μm

Figure 5 Contour of volume fraction of hematite and silica for TC-e160 with inlet, underflow, and overflow taken into account

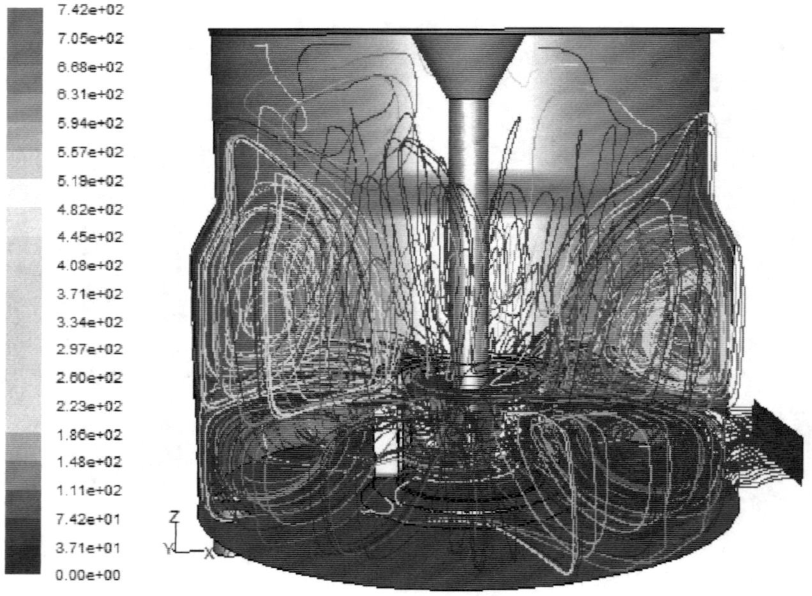

Figure 6 Pathline colored by time (s) with inlet, underflow, and overflow taken into account

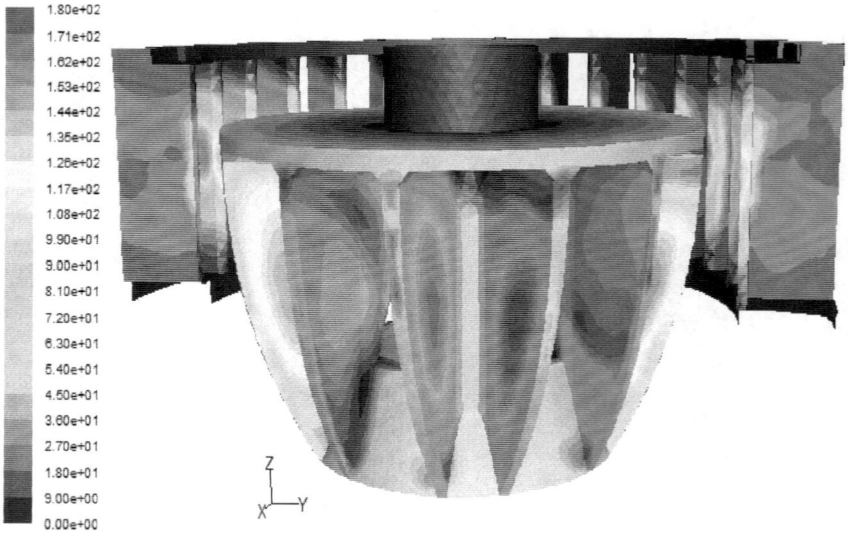

Figure 7 Contour of shear stress on rotr and stator blade surfaces for TC-e300

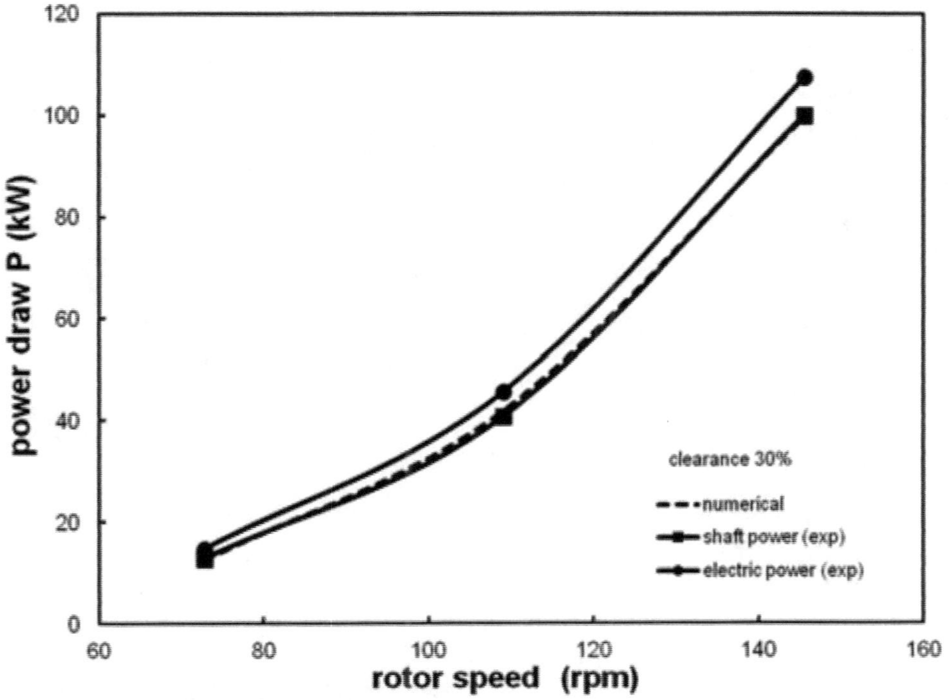

Figure 8 Comparisopn of CFD solution with experimental data from an industrial flotation machine at different rotor speeds for the rotor bottom clearance of 30%

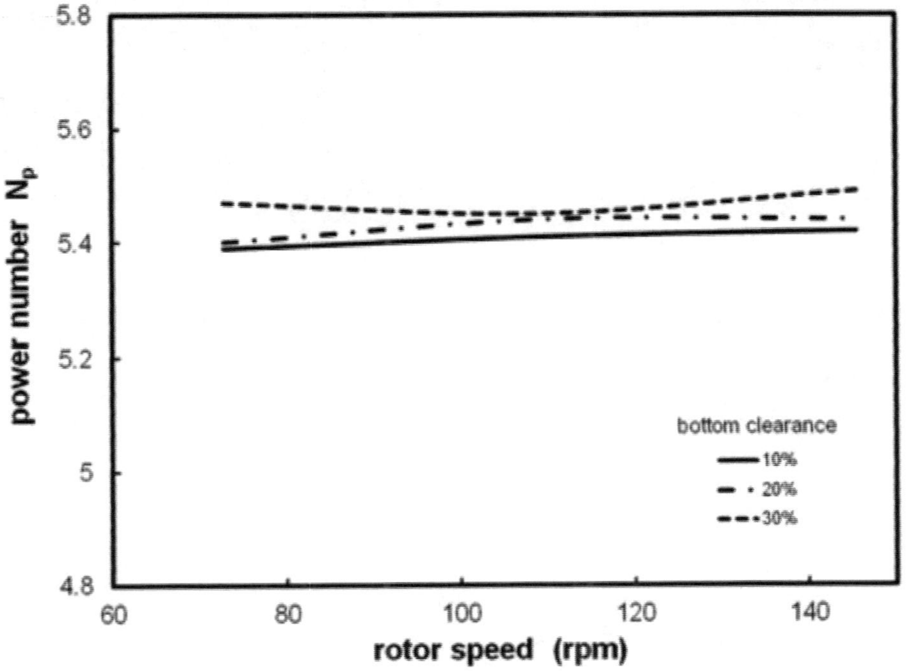

Figure 9 Effect of rotor bottom clearance on power number

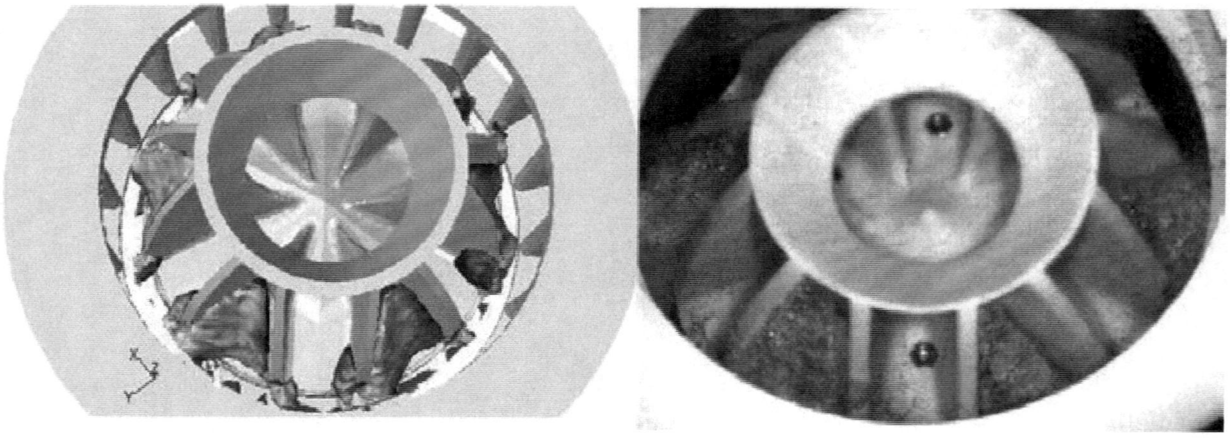

Figure 10 Comparison of CFD solution with experimental observation across the slurry slot and rotor blade slot

Figure 11 Comparison of CFD solution with experimental observation around the stator

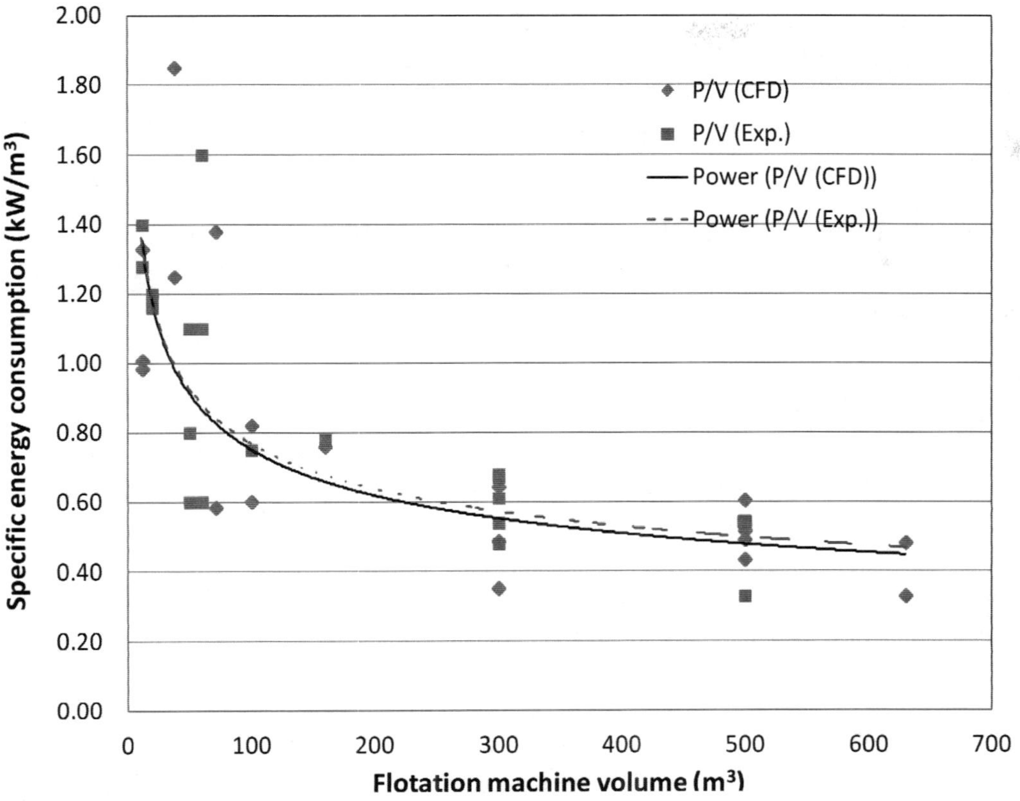

Figure 12 Comparison of specific energy consumption between CFD prediction and experimental measurements for various flotation machine sizes

Summary

The present paper shortly outlines some results of extensive CFD modeling and experimental measurements conducted at Outotec to develop flotation technology. Different flotation mechanisms are examined for varied rotor speed, rotor bottom clearance, flotation machine shape and size, and so on. The CFD modeling has helped to understand flotation mechanism and to keep the Outotec's flotation technology in a leading position. The CFD predictions are in a good agreement with experimental data. As the CFD modeling is conducted without prior knowledge of the experimental results, the mathematical method developed is reasonable to predict flow performance and mixing in flotation machines.

REFERENCES

ANSYS Fluent, 2015.

Chen D., Chen J. H., Fan X. S., and Feng T. R., CFD studies on flow field characteristics of the 600 m3 air-forced mechanical flotation cell, 54th COM 2015.

Grau R., Heiskanen, K., Gas dispersion measurements in a flotation cell, Minerals Engineering, 2003, 16, pp. 1081-1089.

Grau, R A, 2006. An investigation of the effect of physical and chemical variables on bubble generation and coalescence in laboratory scale flotation, Doctor Thesis, Helsinki University of Technology, TKK-ME-DT-4, Espoo.

Koh P. T. L. and Schwarz M. P., CFD modelling of bubble–particle attachments in flotation cells, Minerials Engineering, 9 (6-8), pp. 619-629, 2005.

Laskowski, J S, 2007. Flotation thermodynamics: can we learn anything from it?, Can Metall Quart, 46 (3), pp., 251-258.

Liu, J and Xu, Z, 2007, Role of flotation reagents in tuning colloidal forces for sphalerite-silica separation, Can Metall Quart, 46 (3), pp. 329-340.

Myllykangas, H J, 2009. The effect of bottom clearance on performance of the FloatForce® mechanism, Helsinki University of Techonology Publication, Espoo, Finland.

Sarhan A. R., Naser J. and Brooks G., A review of CFD modeling of flotation cells, 19th Australian Fluid Mechanics Conf., Melbourne, Australia, 8-11 Dec., 2014.

Tatterson, G B, 1991. Fluid mixing and gas dispersion in agitated tanks, McGraw-Hill Education, New York.

Tiitinen, J, Vaarno, J, and Grönstrand, S, 2003. Numerical modeling of an Outopumpu flotation devise, CSIRO 3rd Int. Conf. on Computational fluid dynamics in minerals & process industries, Melbourne.

Xia J. L., Rinne A., Grönstrand S., Effect of turbulence models on prediction of fluid flow in an Outotec flotation cell, Minerals Engineering 22, pp. 880-885, 2009.

Xia J. L., Rinne A., Grönstrand S., Performance of Outotec FloatForce® flotation mechanism – a study in both computational fluid dynamics and industrial scale measurement, XXV Int Mineral Processing Congress -, IMPC2010, Brisbane, Austrlia, 6-10 Sept., 2010

436854 Development, Scale up and Testing of the Nextstep™ – the Most Efficient Mechanism in Mechanical Flotation

Abstract:

FLSmidth introduces the patent pending nextSTEP™ advanced flotation mechanism; the newest design for forced air flotation technology with the lowest power consumption on the market. This new rotor-stator technology is the result of extensive fundamental studies, laboratory test work and full-scale plant testing. Over two hundred different rotor-stator combinations were tested to develop the nextSTEP™ design. This final technology demonstrates significant improvement in both metallurgical performance and energy efficiency.

The exceptional performance of the nextSTEP™ mechanism comes from the rotor and stator being perfectly matched which delivers the best energy dissipation flow, thus maximizing the probability of attachment and flotation. Wide flow jets produced by the nextSTEP™ rotor and the cross pattern from the patented slotted nextSTEP™ stator increase the chances of flotation. Adding slots to the stator increases recovery and boosts overall flotation efficiency. One reason for this improvement is the higher homogenized turbulence dissipation that occurs. In addition to the increased performance derived from using the FLSmidth nextSTEP™ mechanism, it is also relevant to note that the mechanism provides better wear distribution for increased rotor and stator life. This will have a positive impact on the long term operating costs and can benefit both new and existing installations.

424520 Design and Characterization of a Scaled-up Millifluidic Groove Mixer

Abstract:

Mixing in systems at the microscale and macroscale has been well characterized, but less is known about mixing at the millifluidic scale, especially in a laminar regime. Millifluidic systems, defined as having at least one dimension close to 1 mm in size, represent a class of set-ups important to lab scale production of organic compounds[1] and nanoparticles[2]. Millifluidic systems easily have the capability of producing products on the order of grams per day[3], unlike microfluidic set-ups, which due to their inherently small size are unlikely to achieve this level of production. Additionally, specialized fabrication facilities are normally required to manufacture microfluidic devices and typical manufacturing techniques at this scale are not amenable to mass production.

Passive mixers have been well-developed at the microfluidic scale to mix solutions flowing in a laminar regime. In this work we scaled-up the highly utilized groove mixer design[4] to mix solutions in a high throughput manner. We first optimized device dimensions using COMSOL and then fabricated the millifluidic mixer by milling polymethylmethacrylate (PMMA) and thermally bonding the milled PMMA to a flat PMMA substrate. We injected dye or fluorescent solutions into the PMMA mixer and quantified the extent of mixing from brightfield and confocal microscope images. Completeness of mixing depended on flow rates and ranged from ~85 – 95% mixing in ~0.25 – 4 seconds. We also quantified the effects on mixing as a result of injecting inlet streams at unequal flow rates and using solutions with viscosities greater than the viscosity of water. Here we have developed and characterized a simple mixer, amenable to mass production, that has the potential to be used in continuous flow millifluidic systems for synthesizing organic compounds and isotropic or anisotropic nanomaterials.

References

1. V. Dragone, V. Sans, M. H. Rosnes, P. J. Kitson and L. Cronin, *Beilstein J Org Chem*, 2013, 9, 951-959.

2. Y. Li, A. Sanampudi, V. Raji Reddy, S. Biswas, K. Nandakumar, D. Yemane, J. Goettert and C. S. S. R. Kumar, *ChemPhysChem*, 2012, 13, 177-182.

3. S. E. Lohse, J. R. Eller, S. T. Sivapalan, M. R. Plews and C. J. Murphy, *ACS Nano*, 2013, 7, 4135-4150.

4. A. D. Stroock, S. K. W. Dertinger, A. Ajdari, I. Mezic, H. A. Stone and G. M. Whitesides, *Science*, 2002, 295, 647-651.

425252 Scale-up of Solids Suspension - Zwietering or Power per Mass?

Abstract:

Physical and virtual experiments were made on three sized tanks for solid suspensions. The 12”, 20” and 40” tanks had flat bottoms. Hydrofoils of various solidities and pitched blade turbines of several widths were studied in a range of $0.20 < D/T < 0.6$. Water was used as the liquid in all tests and various solids were tested in concentrations from 1% to 40% having particles sizes from 80 to 850 microns. Each of the three tanks were equipped with torque cells and video equipment to determine the just off-bottom speed, 90% suspension and the fully suspended criteria. A virtual mixing (CFD) experiment of each physical experiment was also done to identify the fluid momentum properties at the location of the last suspended particles from the bottom. Results will show comparisons of the two suspension models most commonly used today, Zwietering and power per mass.

This is a continuation of our work that we showed at the North American Mixing Conference in June, 2014 at the Sagamore, Lake George, NY.

425195 Effect of Impeller Submergence on Power Dissipation and Solids Suspension in Mixing Systems Equipped with Pitch-Blade Turbines

Abstract:

Mixing and dispersion of solids in a liquid is a process frequently encountered in the pharmaceutical industry and often conducted in cylindrical baffled tanks stirred by mechanical impellers. In operations where the liquid level is decreased (as often which emptying the tank) the process must be stopped when the solids are no longer suspended. In this work, the minimum agitation to suspend solids (*Njs*) when the liquid level was lowered, and the impeller submergence S_b changed as a result were determined for the case of a six-blade, pitch-blade turbine (6-PBT) impeller. The power consumed by the impeller was also measured. It was found that when a critical impeller submerge level was reached it was impossible to suspend the solids and the impeller power decreased significantly, although the impeller was still full submerged. The results are important to ensure that mixing systems are operated properly.

430972 New Correlation for Predicting Cloud Height in a Stirred Tank

Abstract:

The goal of the study was to obtain a cloud height correlation that included the effects of particle size, particle density, particle concentration, fluid density and viscosity, impeller speed, impeller size and tank diameter. The laboratory scale study was conducted using a Plexiglas, flat bottom tank 28.3 cm diameter tank with a A310® impeller with 8.5 centimeter diameter. The large scale studies were conducted at Philadelphia Mixing Solutions using a 0.61 m diameter tank and and impeller size of 0.2 m. The particles that were suspended included urea formaldehyde, polyoxymethylene (acetal), melamine, sand, and aluminum with specific gravity values of 1.2, 1.4, 1.57, 2.6, and 2.7 respectively. The particle sizes ranged from 0.15 to 9.4 mm. The cloud height was measured at different impeller speeds (measured in RPM) ranging from nearly just suspended speed (NJS) to 2355 RPM, the maximum speed of the equipment. The particle concentrations used within the tank were 1%, 2.5%, and 5% by weight. The viscosity and density of the fluid was varied by using water and solutions of sugar and water.

It was determined that cloud height is dependent primarily on speed and tank size with a lesser dependence on density difference, particle size and viscosity. Concentration has a negligible effect on cloud height.

417240 Vortex Generation, Gas Draw Down and Mass Transfer in Agitated Vessels

Abstract:

Many chemical reactions require the drawdown and incorporation of gas from the head space of an agitated vessel into the liquid where the gas is available for reaction. Examples of these reactions include hydrogenations and nitrations.

There have been many studies relating the development of the vortex to the system vessel geometry and the impeller Froude number but this gives no information about the gas-liquid mass transfer rate which, ultimately, is needed for process design.

Giacomelli and Wyczalkowski (AIChE Annual Meeting, 2013) presented a definition of the minimum Froude number (Fr_{MIN}), when a stable vortex is drawn down to the level of the impeller, and a produced a correlation relating this Froude number to the ratios of the impeller submergence below the unagitated surface (S) and the impeller diameter (D) to the vessel diameter (T):

$$Fr_{MIN} = \alpha \, (S/T)^{\beta} \, (D/T)^{\gamma}$$

The constants α, β and γ were determined experimentally.

Mass transfer coefficients (k_La) have been measured in agitated vessels operating with $Fr \geq Fr_{MIN}$. On scale-up at equal Fr, k_La increases and the reason for this is that, at equal Fr, the power input by the agitator per mass of fluid increases with scale. This generates smaller bubbles and higher interfacial area for mass transfer.

429335 Modeling of Droplet Breakup in a Turbulent Flow

Abstract:

A model of droplet breakup in a turbulent flow is developed. The energy approach is employed: it is assumed that breakup occurs when the kinetic energy of a turbulence eddy interacting with a droplet exceeds a critical value proportional to an increase in the surface energy of newly formed droplets over the surface energy of an original droplet. The breakup event is assumed to be binary. Turbulence eddies are considered as spheres, whose velocities and concentrations are obtained based on the Kolmogoroff theory of isotropic turbulence. However, in contrast to the known models also based on the energy approach, the minimum scale of a turbulent eddy that is capable to break a droplet of a known size into fragments with a given volume ratio is not assumed but calculated. Also, a model of eddy - droplet interactions, different from that employed by majority of modelers, is developed. We suggest an approach, according to which the frequency of these interactions is calculated as a fluid fluctuation frequency, caused by eddies of different scales, in a given point of a computational domain. This approach is free of the strong conventionally used assumption, according to which a turbulent eddy at a moment preceding a collision with a droplet is considered as a sphere moving as whole with a certain velocity. Equations for both the droplet breakup rate and the daughter droplet size distribution are derived. The droplet size distribution function contains a single empirical parameter that is identified using an experimentally validated correlation for the maximum steady-state droplet size of stable non-coalescing droplet dispersion in a turbulent pipe flow. An equation for the droplet breakup rate also contains a parameter that should be identified from experimental data. A computed droplet size distribution is compared with experimental data available in literature.

434202 A Semi-Empirical Approach to Predict Solids Stabilized Emulsion Size

Abstract:

The droplet size of concentrated solids stabilized emulsions produced in an agitated reactor is modelled using a semi-empirical approach. The modelling procedure estimates the coverage and the interface generation potential and assumes that the theoretical stabilized interface is equal to the lowest potential. The coverage potential is obtained by considering the solids properties while the interface generation is deduced from the correlation developed by **Calabrese R.V. et al. (1986)**. The initial Calabrese correlation was modified to include the coalescence effect due to the oil concentrations of up to 20% considered. The effect of particles on both breakage and coalescence processes is also taken into account. Assuming that stabilization by particles is achieved by covering the droplets with a particles network, the stabilization is analysed at two scales: the droplet and the particle. Four steps are used to model particles absorption and four efficiencies are defined. At the particle scale, the particle/droplet attachment should be achieved concomitantly with the collision, then the three phase contact line should be formed and finally the attachment force should be strong enough to avoid detachment. At the d droplet scale the coverage rate should be higher than the coalescence rate. Combining the analysis at the two scales, a good agreement with experimental results is obtained for a wide range of viscosities, for many impeller speeds and solids concentrations of up to 20%. An adapted form of the Calabrese correlation is finally proposed to account for a wider range of conditions than initially and include the presence of solids as stabilizers. The role of the four efficiencies on the modelling approach is highlighted and the impact of operating conditions is emphasised.

421106 Mechanistic Study of Water/Solids Settling in a Diluted Bitumen System: Statistical and Image Analysis

Abstract:

Mechanistic study of Water/Solids Settling in a Diluted Bitumen System: Statistical and Image analysis

Nitin Arora[1] (*nitin@ualberta.ca*)

Suzanne M. Kresta[1] (*kresta@ualberta.ca*)

Samson Ng[2] **(ng.samson@syncrude.com)**

Sujit Bhattacharya[2] **(bhattacharya.sujit@syncrude.com)**

1. *Department of Chemical and Materials Engineering, University of Alberta, Edmonton, AB T6G 2V4, Canada*

2. *Syncrude Canada Ltd., Edmonton, AB T6N 1H4, Canada*

Keywords: *nearest neighbor analysis, froth, sweep flocculation, clays, floc size*

Abstract

The removal of water and solids (clays and heavy minerals) from bitumen froth is a key step in bitumen extraction from Athabasca oil sands [1]. Bitumen froth contains approximately 60% bitumen, 30% water and 10% solids by mass [1]. Bitumen froth is diluted with naphthenic solvent followed by demulsifier addition to remove water and solids. To achieve the desired level of dewatering, process specifications for both the bulk concentration of demulsifier and the mixing conditions (injection concentration, energy dissipation and mixing time) must be determined[2]. The presence of asphaltenes, naphthenic acids and bi-wettable solids on bitumen-water interface[1] makes water/solids separation from bitumen, a slow and difficult process. The goal of this work is to determine the settling mechanism (sweep flocculation, coalescence) and interactions between particles and drops (water-water, water-solid and solid-solid) using spatial statistics and image analysis.

An image analysis protocol for drop size distribution of spherical water drops was previously developed for diluted bitumen system [3]. However, the protocol could not be applied to bitumen froth because of non-spherical water drops and higher solids content (Figure 1). There is a need for development of image analysis algorithm for froth.

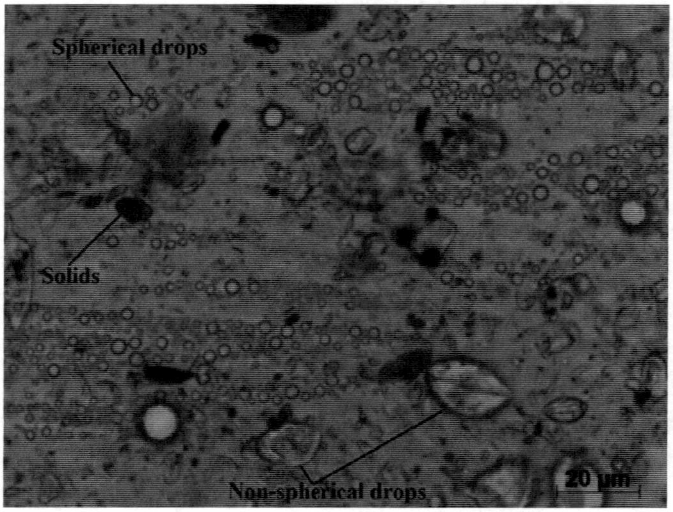

Figure 1: Microscope image with 40X objective containing spherical and non-spherical drops and solids.

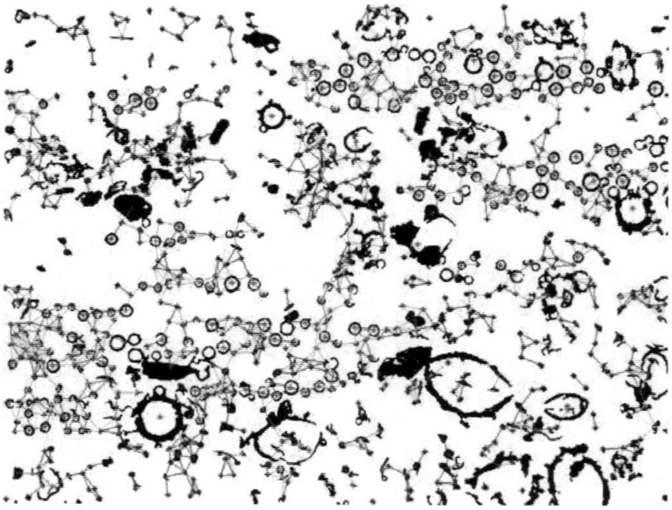

Figure 2: Cluster identification after pre-processing of Figure 1.

We have developed a protocol which isolates objects (solids and water) from a microscope image and determines their location. A series of test images containing two objects (A and B corresponding to water and solids) with varying amounts of clustering have been developed to test the point-nearest-neighbor (PNN) method for cluster detection. PNN method provides an overall clustering tendency in an image; however, the method cannot quantify the number, size and location of clusters in an image space. A separate protocol is being developed to obtain this information (Figure 2). Once developed, the image analysis protocol will be integrated with a real time imaging system to monitor clustering and hence detecting operational problems.

Highlights

- Demulsifier performance can be monitored by cluster detection of bitumen froth images.
- The composition of the clusters: clay-water, water-water, clay-clay, can also be detected.
- The water/solid settling mechanism affects demulsifier performance and settling time.

References

1. Masliyah, J.; Xu Z.; Dabros M ;Czarneck. J. Chapter 7. Froth Treatment, Tipman R. In: *Handbook on Theory and Practice of Bitumen Recovery from Athabasca Oil Sands Volume II: Industrial Practice*. Kingsley Knowledge Publishing; 2013:211-250.

2. Chong JY. Mixing Effects on Chemical Demulsifier Performance in Diluted Bitumen and Froth. 2013.

3. Leo SS. Measurement and Analysis of Changes in Drop Size Distribution during Bitumen Clarification using Image Analysis. 2013.

432430 Mechanistic Model Development of Amine Hydrochloride Salts Production in a Confined Impinging Jet Reactor

Abstract:

The synthesis of polyurethane involves a series of undesired side reactions that result in precipitation of highly insoluble amine hydrochloride (AHC) salts. The production of these salts means loss of starting material and costly reprocessing. Using joint experimental and theoretical studies it is necessary to get a better understanding of these reactions, and identify conditions when the formation of precipitates is either mitigated or the size of the particles being produced is negligibly small.

The formation of AHC salts during the precipitation reaction between 4,4'-methylene dianiline (MDA) and hydrogen chloride (HCl) has been studied experimentally in a confined impinging jet reactor (CIJR) [N. F. Ershad 2013. *Concentration and mixing effects on the production of amine hydrochloride salts in a confined impinging jet reactor.* MSc thesis, University of Alberta, Edmonton]. A mechanistic numerical model that describes the experimental processes in CIJR is now developed. The model involves three simultaneously occurring phenomena: chemical reactions, crystallization, and mixing of reacting species and precipitated particles. Chemical reactions lead to the supersaturation buildup. This supersaturation is relieved by particle nucleation and growth. Agglomeration of the particles due to collisions resulting from Brownian motion and shear-induced collisions is also taken into account. As soon as the particles appear, the transient behaviour of the particle size distribution (PSD) is tracked by solution of the population balance equations. The equations governing reaction crystallization and PSD evolution are coupled with a mixing model that incorporates meso and micro mixing effects in the CIJR. Within the mixing model, the reactor is split into several interacting zones characterized by different values of the energy dissipation rate, entrainment from the surroundings, and residence times.

The empirical parameters of the model are obtained by fitting the numerical results and experimental observations in terms of the size of the particles. The unknown kinetics of the chemical reactions are determined such that the simulations mimic the following experimental result: the particles can be either methylene dianiline monohydrochloride MDA.HCl or methylene dianiline dihydrochloride MDA.2HCl. The particle species strongly depend on the local concentration of amine (the MDA blend strength). The model allows us to investigate the effect of local HCl concentration (percent excess of HCl) and mixing intensity (injection flow rate) on the size and composition of the particles and develop a guideline on how to achieve and control a desired PSD of a certain salt type throughout the process.

433331 CFD Simulation of Solids Supension at High Loading

Abstract:

Predicting the power consumption during solid suspensions remains of key importance in many chemical and Oil & gas related stirred vessel designs. Such predictive capability is essential in scale-up of vessels, especially when the solids loading is high. Multiphase CFD-simulations are employed to gain such predictive capability.

In this work, power consumption for a varying degree solids (concentration of solids) suspended in a benchtop stirred vessel was calculated from CFD simulations with commercial code STAR-CCM+. A polyhedral mesh with prism layers was used. The solids were modeled using the granular pressure and granular temperature transport model in the Eulerian multiphase framework.

Details of the models used and mesh configurations along with results will be presented. The results indicated promising utilization of CFD in design and scale-up of multiphase systems.

428123 Modelling the Hydrodynamics, Transport and Reactions in Multiphase Microreactors

Monday, November 9, 2015: 4:18 PM
Salon H (Salt Lake Marriott Downtown at City Creek)
Lu Yang, Chemical Engineering, MIT, Cambridge, MA

Abstract for the AIChE 2015 Annual Conference

Modelling the Hydrodynamics, Transport and Reactions in Multiphase Microreactors

Lu Yang, Yanxiang Shi, and Klavs F. Jensen

Department of Chemical Engineering, MIT, 77 Massachusetts Avenue, Cambridge MA 02139

On-chip flow chemistry synthesis has advanced rapidly in recent years as a fast and effective means to discover and screen suitable reaction candidates for continuous pharmaceutical production. Among the many chemical reactions, multiphase reactions constitute a major category with important industrial applications. The extent of multiphase reactions is often limited by the innate mass transfer resistance across phase boundaries, and microreactors have been shown to effectively enhance the rate of mixing, therefore improving the efficiency of such reactions. However, compared to single-phase flow chemistry systems, many unknowns remain in the design, optimization and scale-up of multiphase microreactors – primarily due to the complex nature of the multiphase flow

Towards the goal of increasing understanding of the mass transfer enhancement of multiphase flows in microreactors, we have studied hydrodynamics, transport and reactions processes in the segmented flow in an open channel and in a microreactor filled with posts. High-resolution simulations were performed using a series of C++ solvers developed in OpenFOAM. To accelerate computing, the solvers were designed to run in parallel on a high performance computing cluster. Computational fluid dynamic (CFD) simulations of multiphase flow using the volume-of-fluid (VOF) method were in good agreement with experimental observations (Figure 1) and analytic solutions. Based on the validated hydrodynamic solver, we introduced a scalar transport equation with sink/source terms using the one-fluid formulation, which facilitated the simultaneous capturing of multiphase hydrodynamics, mass transfer and reactions. In tandem with the numerical simulations, we also performed mass transfer analysis of multiphase flows based on the penetration theory and a two-stage theory, which further examined the mechanism of mixing enhancement in microreactors, and revealed a two-fold increase in mass transfer coefficients in the post microreactors compared to open channels. The deepened physical understanding of the mixing processes in multiphase microreactors enabled predictions of reactor performance, and the same CFD-based strategy can be readily applied to study other types of microreactor configurations.

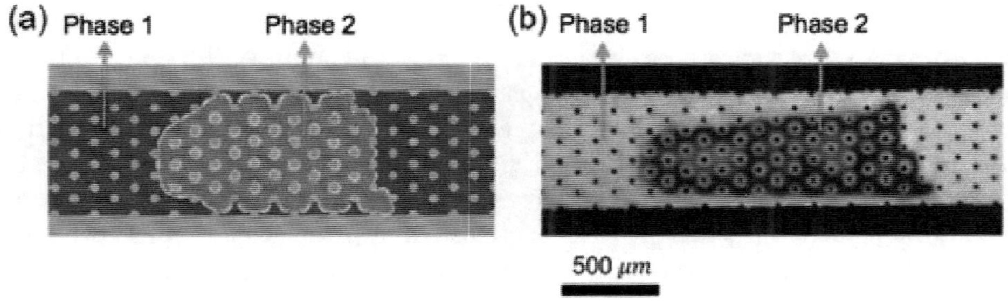

Figure 1 (a) Computational fluid dynamic (CFD) simulation and (b) laser-induced fluorescence (LIF) visualization of segmented flow in the post microreactor. The two immiscible phases can be either gas/liquid or liquid/liquid.

435503 Design of a Pilot-Scale Ebullating-Bed Reactor for Catalyst Testing Aided By CFD and 3D Printing

Abstract:

In ebullating-bed reactors, a bed of catalyst particles is set in weak suspension by an upward flow of liquid and gas reactants. Comparing to fixed beds, the fluidization of the catalyst particles leads to better mixing degrees, which enhances the mass and heat transfer rates, and prevents the bed plugging and channeling due to the formation of coke in certain processes. In addition to better transport properties, the ebullating-bed technology can also grant operation in longer cycles on what catalyst deactivation is concerned: the fluidization allows for a continuous partial removal of deactivated catalyst particles from the reactor and addition of regenerated catalyst. The catalysts usually used in this kind of devices are extrudates with diameters of the order of 1 mm and length around 3-5 mm.

Catalysts in development for these kind of processes must be tested and screened beforehand in its commercial form in smaller pilot-scale units, which try to reproduce the conditions found under the industrial operation. The smaller size of these units allows for easier, faster and less expensive experimental testing by using a smaller amount of catalyst and reactants. However, these pilot-scale units must be quite versatile, enabling testing a large range of conditions: different fluids (gas and liquid charges) and catalysts, and under a large range of flow rates.

In this work, Computational Fluid Dynamics (CFD) has been used to design a new pilot unit based in an internal turbine for the liquid reactive charge recirculation and mixing, and promoting simultaneously the suspension of the catalyst particles. The reactor internal parts have been designed to avoid problems detected in previous existing testing units, such as the flooding of the pumping turbine and the erosion of the catalyst particles. Rapid prototyping of the internal parts of the new unit has been made for experimental validation using the innovative technique of 3D printing. The degree of fluidization of the particles has been quantified from visualization experiments. The degree of mixing in the reactor has been characterized by RTD and mixing time experiments, and from CFD simulations coupled with transport equations of the moments of the age of the fluid.

432223 Removal of Particulate Fines from Organic Solvents Using Water Droplets As Collector Particles

Abstract:

Owing to the pressing need for a reduction in the usage of fresh water in the Clark process of bitumen extraction from oil sands, the oil sands industry has focused its attention on the development of an alternative, solvent-based extraction method in recent years. A key challenge in this new process is the removal of particulate fines from bitumen-solvent mixture. Current methods for particulate removal, such as centrifugation and solvent evaporation, are energy-intensive, and carry potential emission hazards. Motivated by the need for energy efficient methods, our project aims to provide insight into the feasibility of the use of an aqueous phase for collecting and separating fines from the bitumen- solvent mixture by obtaining statistical data for particle - drop impingement process. In this work, a PDMS-based microfluidic aspiration device has been developed to carry out particle-drop impingement experiments. Simulations were performed to obtain a measure of particle interaction time with the drop surface. We observed in the experiments the efficiency of capture of particles by the water droplet *increases* with the impingement rate, a result that is in contradiction to hitherto particle capture efficiency literature. The possibility of the influence of hydrodynamic processes, such as film drainage, and thermodynamically-driven processes such as hole - nucleation and contact line motion, will be discussed. We map out the relationship between particle capture efficiency and flow rate, viscosity ratio and radius of the collector droplets. These results will provide guidelines for the optimal operating parameters to effect efficient particle removal.

437762 Multiphase Application of 3D Printing for Liquid-Liquid Extractor Design

Abstract:

3D printing enables a paradigm shift in the design of complex equipment, making cost-effective the fabrication of a broad range of unique design features which can produce specific flow characteristics and deliver performance enhancement. Liquid-liquid extraction devices such as annular centrifugal contactors must balance their dual roles as both a mixer and separator while also taking into account the residence time requirements of the target chemical process and maximizing throughput. 3D printing provides unprecedented design freedom in meeting these opposing goals with the added potential benefit of lower fabrication costs and greater flexibility in materials of construction.

As guided by insight from multiphase CFD studies and experiments conducted over the last few years, a number of novel centrifugal contactor housing designs have been proposed with the aim of expanding the operational envelope of these devices for application to challenging chemical processes which require increased mixing and residence time to produce efficient separation of desired constituents. One such process under development is for the mutual separation of lanthanides and actinides as part of an advanced used nuclear fuel extraction scheme under development within the US DOE Nuclear Fuel Cycle Technology Program.

Our experience with rapid prototyping and functional testing of a number of custom centrifugal contactor designs of various equipment scales and using real process solutions will be presented. Quantitative comparison of various design alternatives will be made based on evaluation of relative holdup and mixing performance.

Further information on 3D Printing can be found at https://sites.google.com/site/kwardleweb/home

437045 Effect of Impeller Type, Diameter, and Number on Turbulent Blend Time

Abstract:

There are two widely used correlations for predicting turbulent blend time in agitated systems. The approach developed by FMP[1] concludes that all impellers provide the same blend time when compared at the same diameter and power input per mass, while the approach of Fasano and Penney[2], building upon the work of Khang and Levenspiel[3], describes blending as a first-order process governed by a mixing rate constant that is characteristic of each impeller. Data taken with ten impeller styles indicate that both approaches generally provide good turbulent blend time estimates, but do have limitations. The impellers studied in this work include both axial-flow and radial-flow impellers, with turbulent power numbers that span a factor of twenty-five range. In addition to considering the blending performance of single impellers in a square batch (liquid level equal to tank diameter), data is reported for the industrially significant case of multiple impellers in taller batches. The aim of this portion of the study is to understand the effects of liquid level and number of impellers on turbulent blend time, how these parameters interact, and to determine when it is economically justified to add a second impeller.

References:

1. Grenville, R. K., and A. W. Nienow, "Blending of Miscible Liquids", Chapter 9 in Handbook of Industrial Mixing: Science and Practice, edited by E. L. Paul, V. A. Atiemo-Obeng, and S. M. Kresta, Wiley-Interscience, Hoboken, New Jersey (2004).

2. Fasano, J. B., and W. R. Penney, "Avoid Blending Mix-Ups", Chemical Engineering Progress, pages 56-63 (October, 1991).

3. Khang, S. J., and O. Levenspiel, "The Mixing-Rate Number for Agitator Stirred Tanks", Chemical Engineering, pages 141-143 (October, 1976).

422004 The Effect of Viscosity on Stable Vortex Formation in an Unbaffled Stirred Tank

Abstract:

In an unbaffled stirred tank, rotation of the fluid induced by the impeller can lead to the formation of a stable vortex, where the fluid in the centre of the tank is depressed due to centrifugal forces. Depending on process objectives, vortex formation can be desirable or undesirable. If the vortex is large enough to touch the impeller, there can be a loss of mixing power, unwanted gas entrainment, foaming, vibrations, and mechanical stresses on agitator, shaft, and bearings. However under other circumstances it is desirable to form a deep stable vortex – such as to incorporate floating solids or a light secondary phase. Prediction of the vortex depth is, therefore, of practical value in designing an agitation system.

In this work the effect of fluid viscosity on the formation of a stable surface vortex was studied experimentally. Fluid viscosity, impeller speed, impeller diameter, and impeller submergence were varied for a 100L vessel with a Rushton Disk Turbine impeller.

The existing theory for vortex formation (Nagata (1975)) suggests that impeller Froude number adequately characterizes the vortex depth when viscous effects are minor. In the present work, we first show that at the 100L scale, the scaling with Froude number fails when viscous effects are significant. A correlation is proposed to describe the experimental data. This correlation incorporates the effects of Reynolds number and impeller submergence, in addition to the Froude number. This correlation is then validated against data taken at different tank scales.

431836 Effective Transitional Flow Blending

Abstract:

Effective Transitional Flow Blending

In many industrial applications, effective blending in the transitional flow regime is very important to accomplishing process objectives (such as proper reaction kinetic, product uniformity and heat transfer). The ability to effectively blend fluids at lower power and torque inputs offers initial (capital) and longer term (power and maintenance) savings. Many different impeller configurations have been used in the past to accomplish this type of mixing.

Blending in transitional flow has been accomplished in a number of different ways historically. This includes the use of helical ribbons and conventional impellers including straight blade turbines, pitch blade turbines and hydrofoil high efficiency impellers. Although they may provide the best performance in laminar and low transitional operation, helical ribbons are typically not used for transitional blending due to their high cost. To approximate the axial flow pattern and movement throughout the entire tank volume provided by a helical ribbon, multiple open impellers are typically used in transitional operation. The design for effective blending using this approach is a function of fluid rheology and Reynolds number. The number of impellers required to mix the entire vessel increases with both a decrease in Reynolds number and an increase in shear rate sensitivity of a material. The impeller D/T increases with both a decrease in Reynolds number and an increase in shear rate sensitivity of a material.

This presentation will introduce a new transitional flow impeller that provides significant improvements over existing designs. A comparison to existing impeller technologies will made over the Reynolds number range for transitional flows (Reynolds numbers between roughly 100 and 1000). A range of D/Ts will also be examined. Power numbers and blend times will be reported. In addition, a map of where different impeller technologies are optimal will be presented.

430176 Towards a General Transport Equation for Mean Age Distribution

Abstract:

There are several formats of transport equations for mean age distribution are discussed in literature, with different method of derivations. The method discussed by Liu and Tilton is connected to the existing residence time distribution method widely taught in chemical engineering corriculum. In this talk, these equations and methods willl first be reviewed. A general transportation for mean age, applicable to transient, multi-species, and multiphase flows will then be discussed.

More details of this abstract will come later.

413576 CFD Study for Dispersion of Hydrocarbons Near Flare

Abstract:

Flare stacks are key system in the petrochemical process industries, where off gases containing hydrocarbons are combusted before sending it to atmosphere. This study discuses a molecular seal gas leak upstream of a flare stack in a commercial scale petrochemical plant. This leak is difficult to fix during the continuous plant operation, so it had to wait until next turndown time. To continue operation, risk analysis requires the calculations of the distribution of the leaked gas in the proximity of the flare to rule out potential fire/explosion. Flammability limits (LEL: Lower explosive limit / UEL: Upper explosive limit for each individual species are identified. The diffusivity coefficient of each species in air is determined from available empirical correlations. The combustible hydrocarbon dispersion in the atmosphere from the flare line is predicted using CFD modeling. Computational domain for CFD model consist of the atmosphere surrounding the flare stack near leak. Prevailing wind velocity and direction data is used for the atmospheric domain boundary conditions. A 2D model is used to identify the extent of computational domain of interest for the 3D simulations. The preliminary 2D model results are shown in Figure 1. The extent of the flammable zone for the Methane gas is shown in Figure 1b. 3D simulations will also be presented to capture the effects of the flare stack body on flow turbulence and leak gas distribution including parametric study to assess wind velocity/direction and flare gas composition.

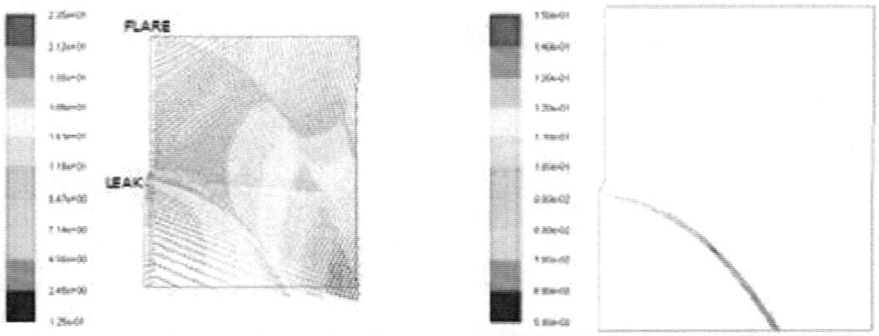

1a) Wind velocity vector (m/s) 1b) Flammability limit of CH4 mole fraction

Figure 1. 2D simulation results for hydrocarbon dispersion in air

440120 Influence of MESH Refinement Near Walls in HEAT Transfer Predictions in Stirred Tanks

Abstract:

Heat transfer in stirred tank reactors (STR) is widely used in the chemical industry. Dimple jackets, halp-pipe jackets and internal coils are employed for both heat addition or removal. The aim of this work is to obtain a CFD model for these systems in order to improve their design. To obtain a reliable CFD model, it is very important to understand the effect of the dimensionless parameter y^+ in the prediction of the heat transfer at the heat exchanging surfaces. This work shows this analysis for a STR with internal helical coils. The validation of the model used the experimental results published by Oldshue and Gretton (1954), which were compared with the CFD results. The numerical CFD data was obtained by the ANSYS CFX 14.0 package. It has been noticed that the variable y^+ did not show any significant effect on the prediction of the primary flows generated by the impeller. However, when it came to the prediction of the heat transfer coefficient, it showed a strong influence and results indicate it is necessary to use values of $y^+ <0,1$ (using the SST turbulence model) to guarantee that the near wall treatment of the data predict correctly the flow.

• Oldshue, J. Y. and Gretton, A. T., "Helical Coil Heat Transfer in Mixing Vessels", Chem. Engng. Progress, V. 50, n° 12, p. 615-621, 1954.

431570 Comparison of PIV Measurements with CFD Simulation Results in an in-Line, Slot and Tooth Rotor-Stator Mixer

Abstract:

Rotor-stator mixers, with their ability to create high shear fields have a broad spectrum of applications in the chemical, petrochemical and pharmaceutical process industries. To better understand device performance and quantify mixing and dispersion capabilities, analyzing the flow field created by the rotor-stator interactions is crucial. We have previously reported 2-D Particle Image Velocity (PIV) measurements, for several rotor speeds and throughput conditions, for an in-line IKA prototype rotor-stator mixer containing single rows of 12 rotor teeth and 14 stator teeth. The working fluid was water in turbulent flow. In this work we report the development and validation of an accurate CFD model and make quantitative comparison between simulation results and the PIV data.

The model geometry and mesh were developed within Ansys Workbench with the fully transient (sliding mesh) 3-D RANS simulations performed with Fluent using the realizable k-epsilon turbulence model. The effect of mesh density and wall treatment were systematically studied to optimize the numerical scheme. With respect to post processing, the numerical data were sampled in a stator slot at 9 rotor tooth positions on a grid that closely mimicked that for PIV data acquisition. Detailed comparisons were made for three different rotor speeds but at the same volumetric throughput.

Our preliminary study of near-wall modelling techniques considered Non-Equilibrium Wall Functions (NEWF) and Enhanced Wall Treatment (EWT). Both produced similar results but EWT showed advantage in computational efficiency. In the mesh independence study, mesh levels of approximately 2, 6, and 16 million cells were considered. The comparison revealed that the model with 6 million cells was sufficient to insure grid independence and solution accuracy.

The results show that the CFD model predicted the location of mixing layer vortices within 12% relative to the stator slot width and accurately captured the general flow movement. However, noticeable differences were observed as well. The CFD model consistently reported higher velocity magnitude (~20% on average) and somewhat different flow direction near the slot exit, where it adjoins the volute region. Despite the differences, the results show that CFD simulations can be used to gain knowledge of flow structure and device performance. The reasons for model and data mismatch are discussed and methods to reduce the mismatch in future studies are suggested.

420709 Large Eddy Simulation of CSTR Systems: Comparison to Theory and Experiment

Abstract:

The mixing properties of CSTR systems are predicted using fully transient large-eddy lattice-Boltzmann simulations. By adding massless tracer particles to the tank inlet, the spatiotemporal variation in the mean-age of the fluid and the local fluid residence time can be calculated on-the-fly throughout the course of the simulation. In addition to identifying the mixing behavior of well-stirred systems, this modeling approach can be used to predict the behavior of imperfect CSTR systems, wherein some portion of the particles leave the system prematurely. Predictions from these simulations are compared to predictions from analytical models and to experimental data. Recommended relationships between grid resolution, tracer particle concentration, and simulation runtime are presented. The extensibility of this approach to study tracer particles with mass (and inertia), along with its ability to predict local eddy dissipation rates and transient impeller blade forces is also discussed.

428282 Computational Fluid Dynamics Simulation of a Pipeline Rotor-Stator Mixer

Abstract:

Rotor-stator mixers provide high deformation rates to a relatively limited volume, resulting in a region in which intensive mixing, milling, and/or dispersion/emulsification operations can occur. Because they are geometrically complex, experimental measurement of important flow regions within rotor-stator mixers is often difficult, if not impossible. Computational fluid dynamics (CFD) simulations of mixers are a means to provide flow data that are of benefit to both designers and end users.

This presentation focuses on our recent work in the fully transient (sliding mesh) three-dimensional RANS simulation of a pipeline rotor-stator mixer, the Greerco High Shear Pipeline Mixer, using ANSYS FLUENT. The modeled unit consists of two conical rotor-stator stages, each with a nominal diameter of 4 inches, aligned to produce an axial discharge flow. The realizable k-ε model was used to model turbulence, while wall effects were modelled with enhanced wall treatment. The mixer geometry consists largely of cylindrical and conical bodies; the impact of this on meshing will be discussed. The working fluid was water in turbulent flow.

With a relatively coarse mesh, the predictions of throughput by the FLUENT simulation matched the measured values provided in the Greerco Pipeline Mixer literature. However, a greater mesh density is required to resolve the velocity fields. The effect of mesh density is illustrated by comparing selected velocity profiles. The mesh-independent solution is used to show various flow field features. As with other rotor-stator mixers, fluid is shown to impinge on the trailing surface of the stator slot. In addition, it is shown that the conical shape of the rotor and stator causes flow rates through the outer regions (where the rotor tip speed is higher) to be much higher than at the inner regions. In the case of the second stage stator, very little flow enters the inner two rows of slots. The effect of the flow field predictions on various aspects of device performance will be discussed.

SPXFLOW

R·I·T

Predicting Baffle Forces Using Both CFD and Experimental Techniques for Both Radial and Axial Flow Impellers

Richard Kehn: Graduate Student at Rochester Institute of Technology (M. Eng. Mech. Eng. 2015)
Director – R&D @ SPX Flow Inc. (LIGHTNIN Mixers), Rochester, NY., USA

Dr. Edward Hensel: Associate Dean for Research and Graduate Studies
Kate Gleason College of Engineering
Rochester Institute of Technology, Rochester, NY., USA

AIChE Annual Meeting 2015 – Salt Lake City, Utah

›Lightnin

COMPANY CONFIDENTIAL

A Quick Review of Baffles

SPXFLOW

- Flow-directing vanes or panels in a vessel mounted near the wall
- Convert radial motion into axial motion and prevent the tank's contents from swirling
- Allow impeller to transfer energy into the fluid
- Must be mechanically designed to withstand the force of the swirling fluid
- For large vessels, the cost to add baffles is not trivial

›Lightnin

Baffles are a key component to an optimized mixer design

R·I·T

COMPANY CONFIDENTIAL

November 23, 2015 2

Predicting Baffle Forces

SPXFLOW

- Pm = Agitator Motor Power (HP)
- DF = Dynamic Load Factor
- B = Number of Baffles
- T = Tank Diameter (ft.)
- N = Rotational Speed (rpm)

- CFD can be used to determine pressures acting along a surface, which can be used to predict forces acting on those surfaces
- But are the predictions accurate?

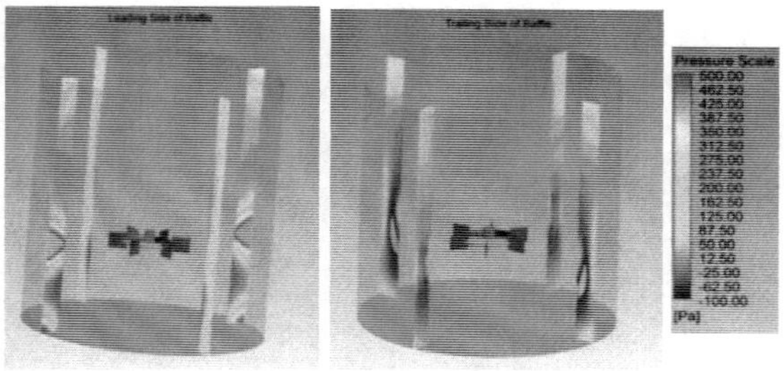

$$F = \frac{10{,}500 \times P_m \times DF}{B \times T \times N}$$

›Lightnin

Correlation courtesy of Oldshue (Fluid Mixing Technology 1983)

R·I·T

Computational Fluid Dynamics (CFD)

SPXFLOW

- CFD (**C**omputational **F**luid **D**ynamics) is the science of predicting fluid flow, heat and mass transfer, chemical reactions and related transport phenomena by solving numerically the required set of governing transport equations

- Developing a model requires the following:

  - Understanding the process you are modeling (e.g. have you seen a tank mix?)
  - Applying the appropriate model to the real life case (is your flow laminar, turbulent, steady state, transient?)
  - Geometry decomposition and mesh techniques (mesh quality) affect the model accuracy and speed of solution
  - Validation of the results?

›Lightnin

CFD is not a stand alone tool....

R·I·T

Problem Statement

SPXFLOW

- Use CFD to predict baffle forces using a steady state approach for two impellers (Rushton vs. PBT)
- Compare results with experimentally obtained results and to the baffle force correlation
- How sensitive are the CFD results to mesh?
- Do both impellers yield the same baffle force at equal power density?

48" Dia. Laser Lab Test Tank
Water as Fluid

16" R100 Rushton Turbine
1D off bottom
90 rpm

16" A200 Pitched
Blade Turbine
1D off bottom 137 rpm

ANSYS R14.0 (FLUENT) used to perform CFD modeling

Geometry / Mesh Generation

SPXFLOW

- Geometry decomposed into 10 separate zones to control cell count near baffle surfaces and to maximize swept meshes throughout model.
- Edge mesh controls used to control mesh sizing on tank edges and baffle edges
- Impeller surfaces controlled with surface sizing function.
- Five meshes were modeled:
    - Four (4) with cell count near baffle varied (cases 1, 1a, 1b, 1c)
    - One (1) with cell count near impeller varied (case 2)

Geometry and Mesh for R100 (A200 similar)

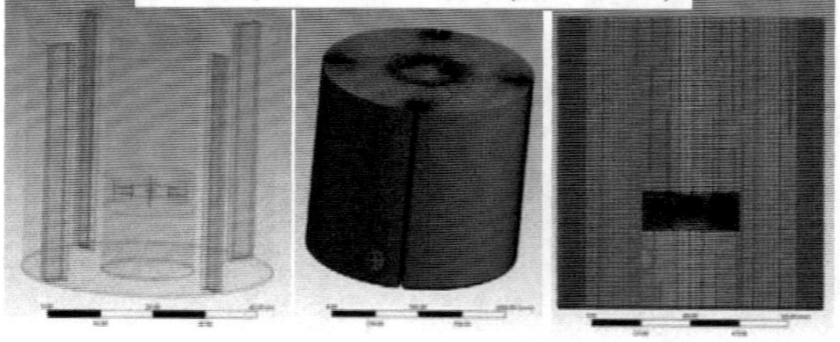

Mesh Near Baffle Surfaces

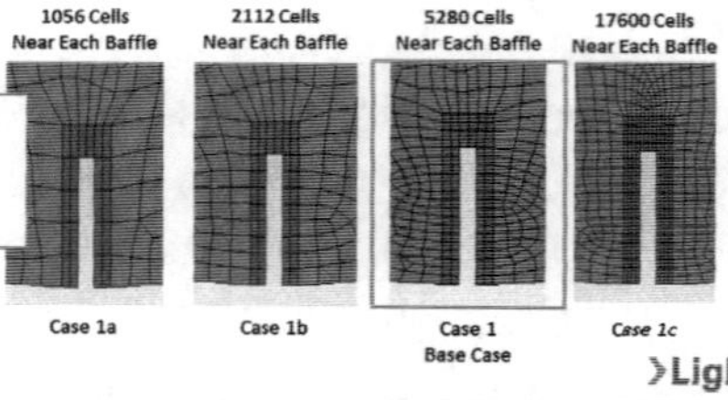

Lightnin

Total number of cells varied between ~400,000 and ~1,600,000 total cells.
Average cell skewness was 0.195 for all cases

Solution Methods

SPXFLOW

- ANSYS Fluent R14.0 used to perform CFD modeling
- Pressure Based Navier Stokes double precision solver used. Second order upwind discretization
- Realizable K-Epsilon Turbulent model used with default settings
- Multiple Reference Frame Steady State Approach used to model impeller rotation
- Convergence typically occurred between 3800 and 4500 iterations. Scaled residuals were below $1E^{-04}$

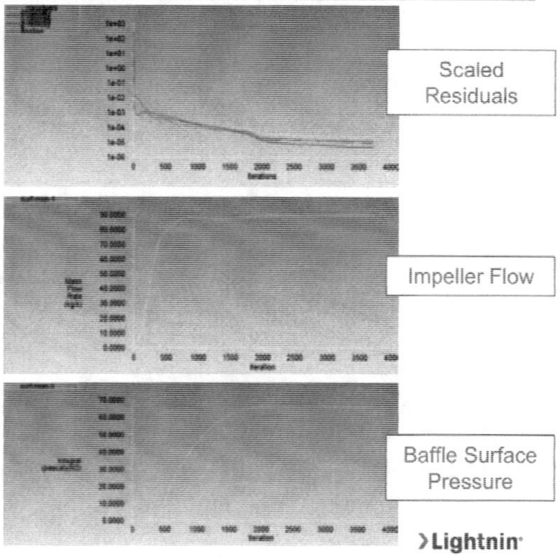

›Lightnin

Monitors on flow, torque and pressures near baffle recorded for each model

Effect of Mesh on Results for R100

SPXFLOW

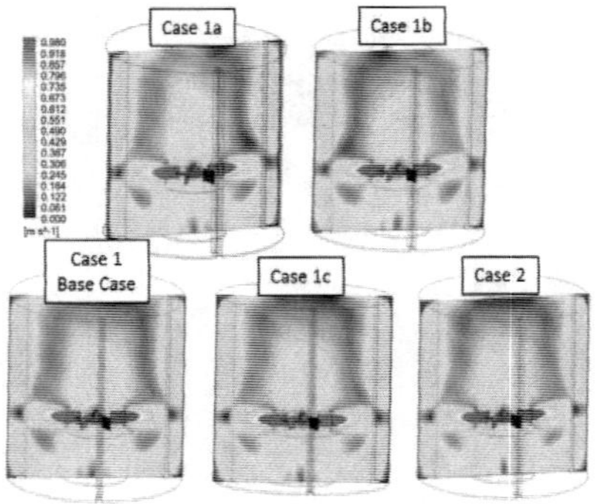

Model Case:	CFD Predicted Np:
Case 1a	4.76
Case 1b	4.76
Case 1 (Base Case)	4.75
Case 1 c	4.73
Case 2	4.78

$$T = \frac{P}{N}$$

Torque

$$N_p = \frac{P}{\rho N^3 D^5}$$

Power Number

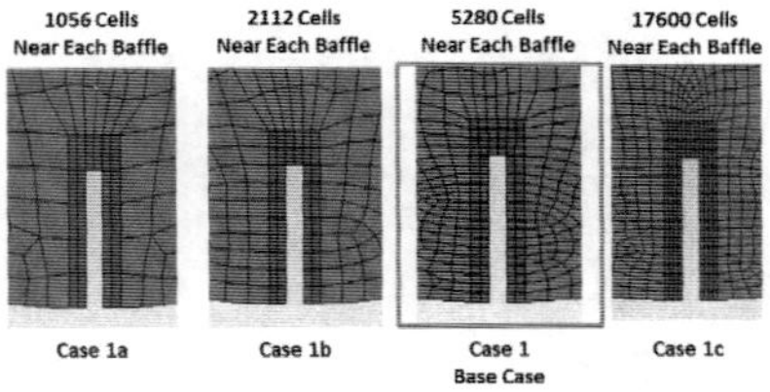

- The flow field and impeller power number predicted is mesh independent
- The power prediction is low (by 8.5%) compared to published and measured power numbers for a Rushton impeller (standard Np = 5.2)

›Lightnin

A200 CFD predicted power number was 1.26 for all simulations

Baffle Force Prediction vs Mesh for R100

SPXFLOW

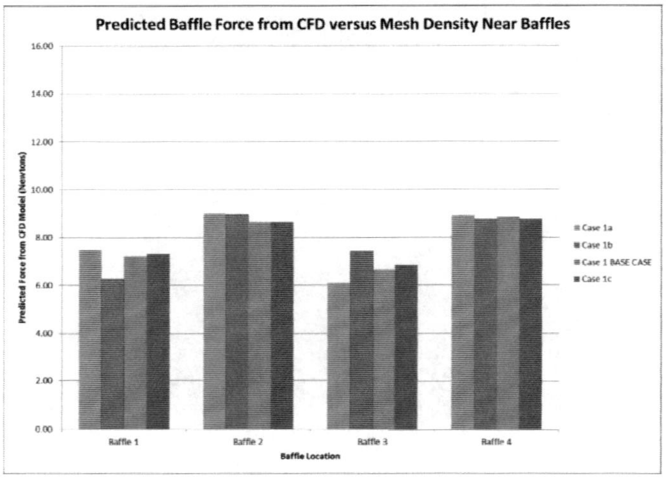

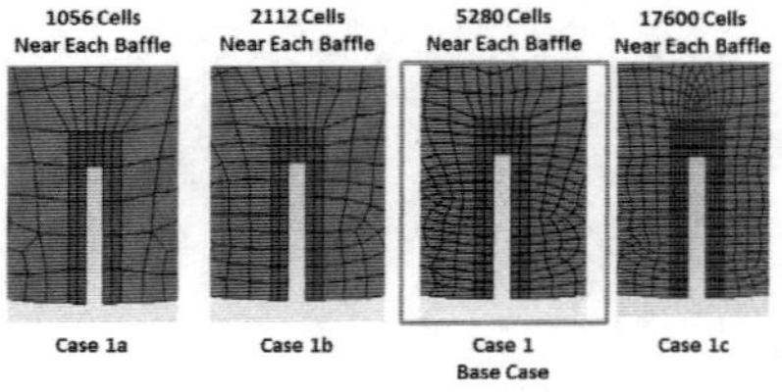

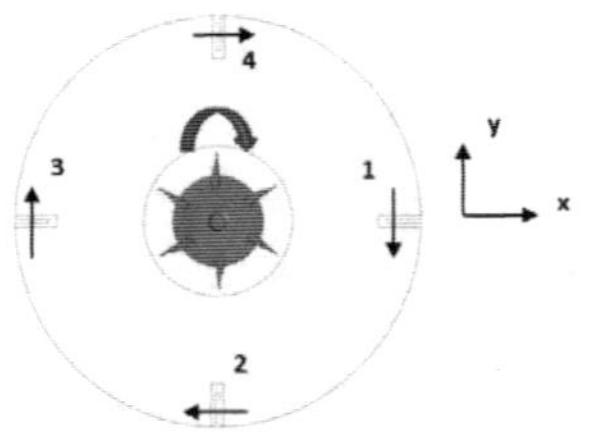

›Lightnin

Case 1 Mesh was chosen to perform the remaining modeling presented herein

Laser Doppler Velocimetry Validation

SPXFLOW

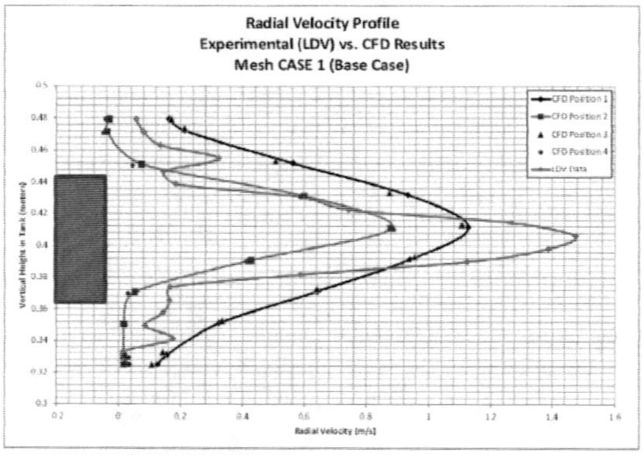

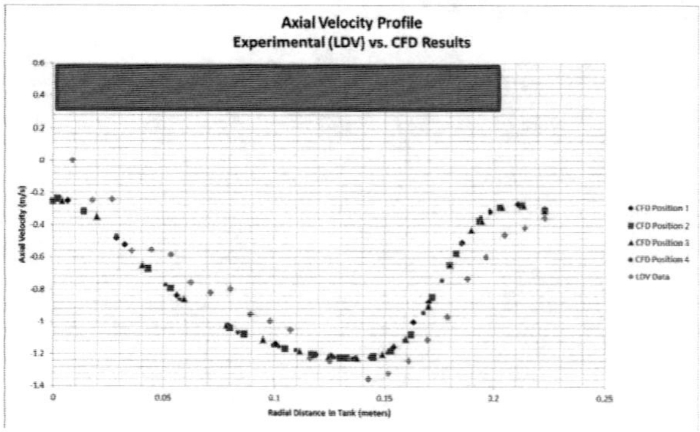

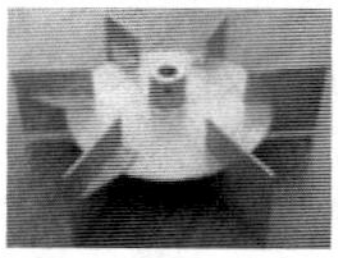

16" R100 Rushton Turbine

16" A200 Pitched Blade Turbine

›Lightnin

Good agreement with LDV data (areas in blue are the impeller locations)

Baffle Force Prediction vs. MRF Location for R100

SPXFLOW

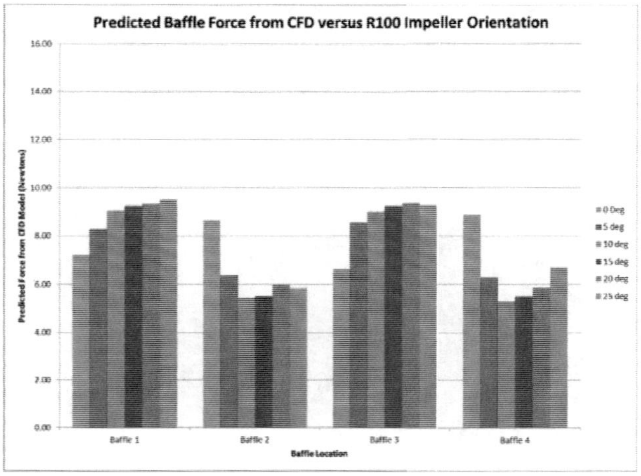

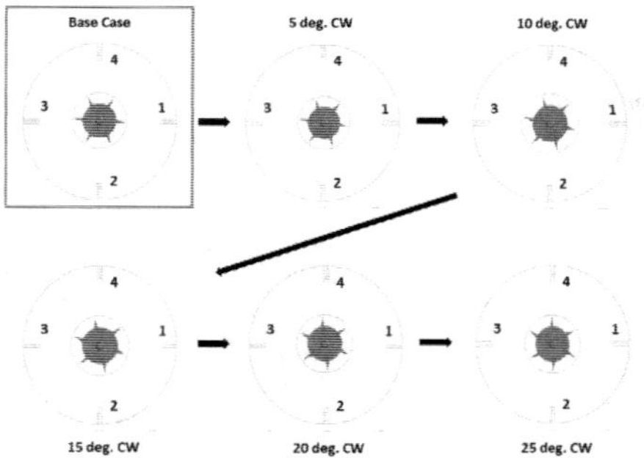

16" R100 Rushton Turbine

Lightnin

MRF Location has a strong effect on predicted baffle force for R100 impeller

COMPANY CONFIDENTIAL November 23, 2015 11

Baffle Force Prediction vs. MRF Location for A200

SPXFLOW

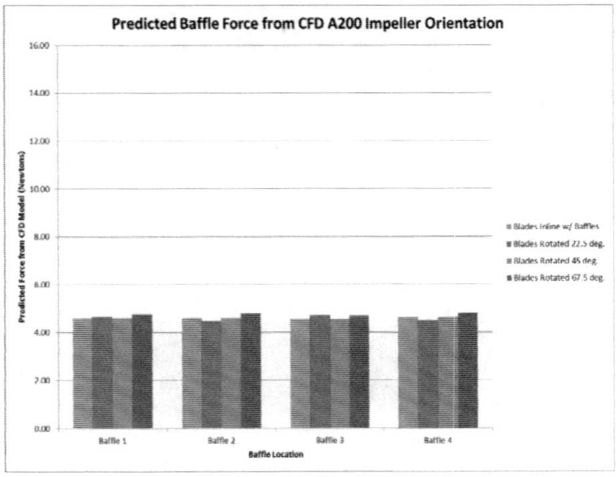

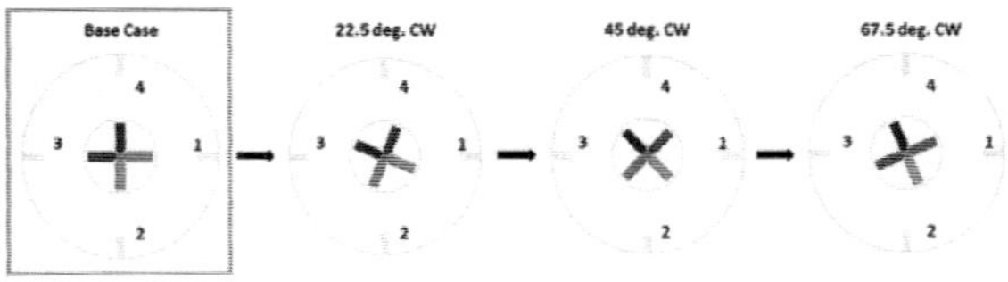

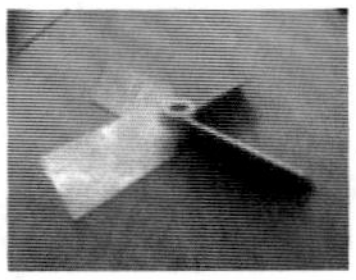

16" A200 Pitched Blade Turbine

Lightnin

MRF Location does not have a strong effect on predicted baffle force for A200 impeller

COMPANY CONFIDENTIAL November 23, 2015 12

Baffle Force Comparison (R100 vs. A200)

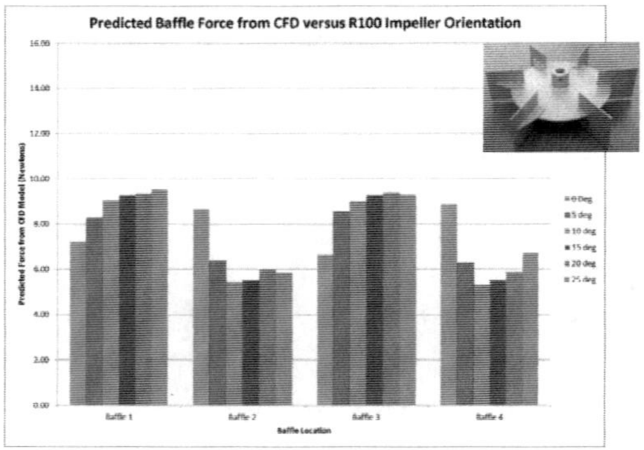

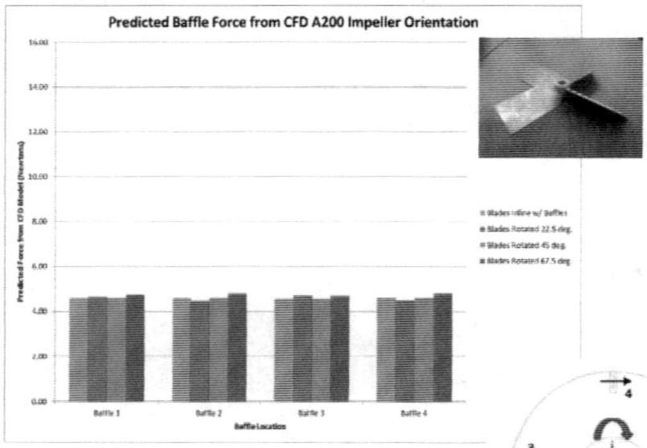

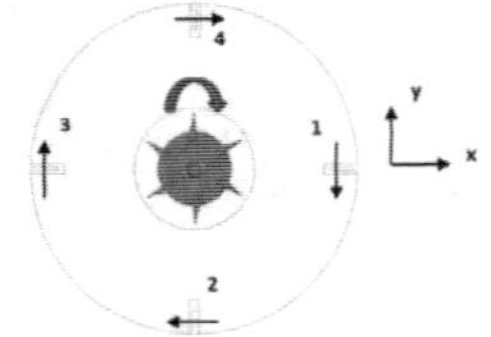

- Maximum baffle force is 9.3 N for the R100 impeller
- Maximum baffle force is 4.6 N for the A200 impeller
- Baffle force equation from Oldshue would predict 15.2 N and 14.4 N respectively for the R100 and A200 (if using the SHP for each impeller versus MHP)

Axial flow impellers impart less force on the baffle than radial flow impellers

Experimental Technique for Estimating Baffle Forces

- Measure deflection of baffle as the mixer is running over the height of the baffle
- Compare this to the deflection measured at various heights with a known force applied (4.45 N, 8.9 N, 13.35 N)
- Observe both average deflection and peak deflection over time (over a 5 minute period)

Experimental Results for Baffle Deflection

SPXFLOW

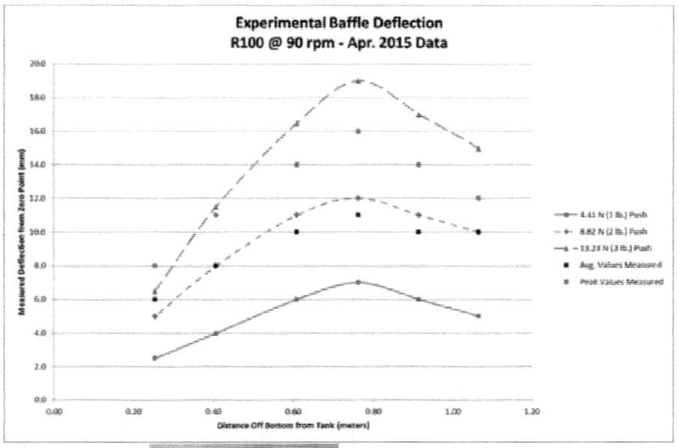

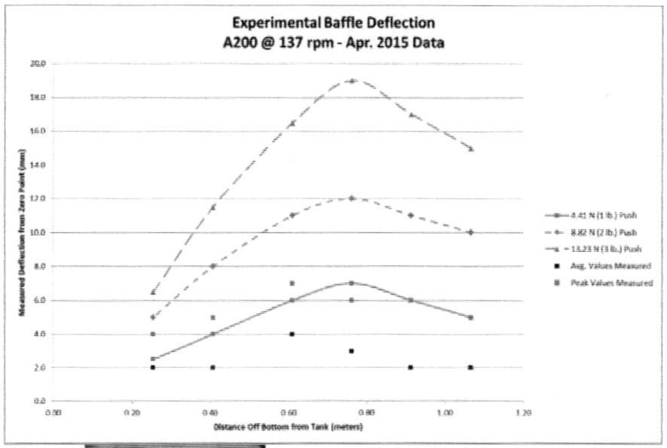

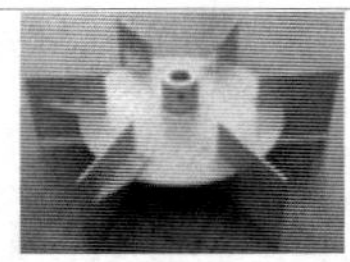

16" R100 Rushton Turbine

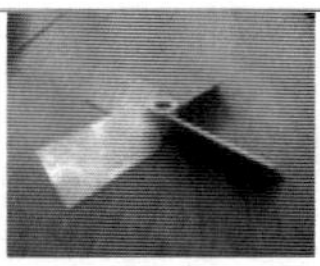

16" A200 Pitched Blade Turbine

Lightnin

Good agreement between CFD prediction and experimental results

Recent Industrial Example

SPXFLOW

Existing System with walls and old baffles

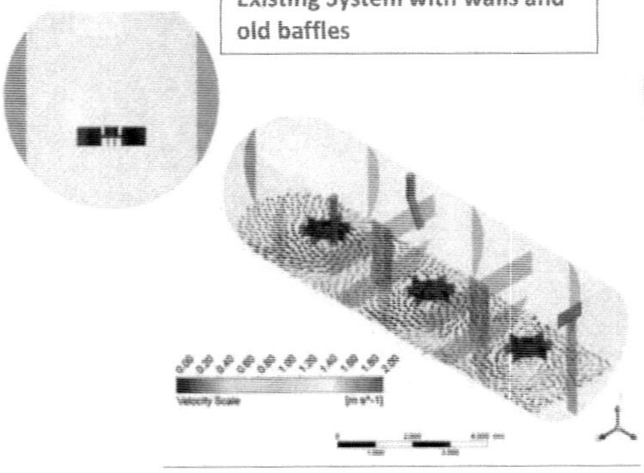

Modified system with walls removed and new baffles

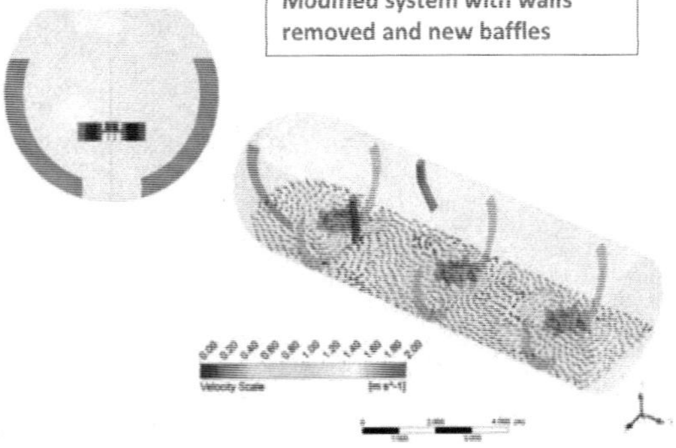

Lightnin

CFD used to predict baffle forces for vessel supplier to design baffles

Conclusions / Future Work

SPXFLOW

- CFD can be used to predict baffle forces for both radial and axial impellers when proper modeling techniques are used. Consideration of blade location relative to the baffle is important when conducting steady state models, especially for radial impellers

- The forces predicted by CFD for these cases matched closely with experimentally measured forces. The traditional baffle force correlation is more appropriate for radial impellers versus axial flow impellers

- Future considerations:

  - Transient model comparison (results vs. time spent modeling). Is the steady state approximation good enough?
  - Geometry effects on baffle forces (multiple impellers, D/T, location)
  - Better experimental validation (strain gauges on metal baffles)
  - Dip tube forces

4 Ft. Tank Test Stand

›Lightnin

namf

R·I·T

COMPANY CONFIDENTIAL

November 23, 2015 17

404589 Modeling Distributed-Parameter Mixing Processes Using Computational Fluid Dynamics and z-Transform

Abstract:

Computational fluid dynamics (CFD) has become a widely used method for analyzing the performance of mixing unit operations in the chemical process industry. Results of these simulations can yield useful insights into a mixing process. Repeated CFD simulations to capture the dynamic behavior of the same mixing process set-up under different sets of boundary conditions can, however, demand significant computational resources and present a practical challenge to practitioners.

We have previously illustrated how the combination of CFD with the z-transform technique provides a convenient approach to analyzing the open-loop response of a lumped-parameter process. Using a continuous stirred-tank reactor as an example, we showed that the simpler z-transform model has excellent accuracy and relatively low computational cost requirements (Yin and Yu, AIChE 2015 Paper 354913). In our current presentation, we extend this methodology to the analysis of distributed-parameter mixing processes where multiple physics and spatial heterogeneity may manifest in the system. One example of such a system is a stirred-tank coupled with heat transfer through the vessel wall. Using the results of an initial CFD simulation we derive a z-transform model which can then be inverted to give the dynamic response of the mixing process for any arbitrary perturbation. We compare the results obtained from the z-transform model against results of CFD simulations to demonstrate the accuracy of this approach and savings in computational cost, and discuss the sources of error and the limitations of this method.

429669 Solution of the Fokker-Planck Equation Using the Extended Quadrature Method of Moments

Abstract:

One of the important attractions of employing the transported probability density function (PDF) methods for turbulent reacting flows is the fact that the chemical source terms are treated exactly. The composition probability density function (PDF) evolves with convective transport in physical space due to the mean velocity (macromixing), the turbulent diffusivity which transports the composition PDF in physical space (mesomixing), the transport in composition space due to molecular mixing (micromixing) and chemical reactions. The key model term of our interest in this study in the transport PDF equation is the molecular mixing term, which describes how molecular diffusion affects both the shape of the PDF and the rate of scalar variance decay. Molecular mixing is expressed using the Fokker-Planck model (Fox, 2003, 1994), which is an extension of the interaction-by-exchange-with-the-mean (IEM) model. The IEM model, widely used in chemical-reaction engineering and computational combustion due to its simple form, assumes the linear relaxation of the scalar towards its mean, while the Fokker-Plank model considers, in addition, the effect of the differential diffusion process to mimic the mixing. Differential diffusion occurs when the molecular diffusivities of the scalar fields are not the same. An extended quadrature-based moment method (EQMOM) (Yuan et al., 2012; Chalons C. Fox R.-O. Massot M., 2010) is used to close the transport equation for the composition PDF. The β kernel density function is used for EQMOM since the scalar composition is represented by a scalar bounded between -1 and 1. The PDF predicted by the EQMOM model in the same conditions studied in the direct numerical simulation of Eswaran and Pope (1988) are reported and compared to the DNS predictions.

References

Chalons, C., Fox, R. O., Massot, M., 2010. A multi-Gaussian quadrature method of moments for gas-particle flows in a LES framework. Studying Turbulence Using Numerical Simulation Databases, Center for Turbulence Research, Summer Program 2010, Stanford University 347–358.

Eswaran, V., Pope, S.B., 1988. Direct numerical simulations of the turbulent mixing of a passive scalar. Physics of Fluids 31, 506–520.

Fox, R.O., 2003. Computational Models for Turbulent Reacting Flows. Cambridge University Press.

Fox, R.O., 1994. Improved Fokker–Planck model for the joint scalar, scalar gradient PDF. Physics of Fluids 6, 334–348.

Yuan, C., Laurent, F., Fox, R.O., 2012. An extended quadrature method of moments for population balance equations. Journal of Aerosol Science 51, 1–23.

429987 A Quadrature-Based CFD Model for Single-Phase Turbulent Reacting Flows

Abstract:

Computational fluid dynamics (CFD) modeling of turbulent reacting flows has a variety of applications, including minimizing the amount of pollutants from internal combustion engines and maximizing the selectivity of a desired product in a chemical reactor. One goal of modeling these flows is to better understand how fluid mixing affects selectivity. A transport equation for the joint composition probability density function (PDF) can be used to model for this purpose (Fox, 2003). Here, the interaction-by-exchange-with-the-mean (IEM) model is used to close the micromixing term in the joint composition PDF transport equation. Both the Conditional Quadrature Method of Moments (CQMOM) (Yuan and Fox, 2011) and the Direct Quadrature Method of Moments (DQMOM) (Fox, 2003) are used in this study to solve the joint PDF transport equation. Compared to direct numerical simulations (DNS), these two methods reduce the computational cost significantly and are applicable to simulate large-scale systems by forcing lower-order moments of a presumed PDF to be exactly preserved (Fox, 2003). Despite reducing the computational cost, statistical methods like these introduce errors that deviate from exact solutions obtained from DNS. In the past, DQMOM has been studied and found comparable results to previous methods (Zucca et al., 2007), but DQMOM introduces correction terms to the transport equations. CQMOM does not introduce these correction terms, and the optimal moments (Fox, 2008) are transported directly rather than as weights and weighted abscissas. Here, a competitive-consecutive reaction system is modeled by the PDF transport equation and solved using DQMOM-IEM and CQMOM-IEM to compare the statistical errors and computational costs of each method. The CQMOM formulation results in a robust solution algorithm for the PDF transport equation with a similar computational cost by avoiding the potentially singular correction terms arising in DQMOM-IEM (Akroyd et al., 2010).

References

Akroyd, J., Smith, A.J., McGlashan, L.R., Kraft, M., 2010. Numerical investigation of DQMOM-IEM as a turbulent reaction closure. Chemical Engineering Science 65, 1915–1924.

Fox, R.O., 2003. Computational Models for Turbulent Reacting Flows. Cambridge University Press.

Fox, R.O., 2008. Optimal moment sets for multivariate direct quadrature method of moments. Industrial & Engineering Chemistry Research 48, 9686–9696.

Yuan, C., Fox, R.O., 2011. Conditional quadrature method of moments for kinetic equations. Journal of Computational Physics 230, 8216–8246.

Zucca, A., Marchisio, D.L., Vanni, M., Barresi, A.A., 2007. Validation of bivariate DQMOM for nanoparticle processes simulation. AIChE Journal 53, 918–931.

432546 Applying Evolutionary Algorithms in Optimizing Mixing in Selective Catalytic Reduction Systems Via Computational Fluid Dynamics (CFD)

Abstract:

One of the foremost applications of computational fluid dynamics modeling techniques over the past ten years has been in the area of optimizing Air Pollution Control (APC) systems used to reduce emissions from fossil fuel fired stationary sources such as power plants and steel mills. As the paper shall demonstrate, the optimization of static mixing devices is critical to the efficient performance of technologies that reduce NOx,. This paper shall describe the authors' experiences in optimizing the fluid flow performance of a Selective Catalytic Reduction (SCR) system that was installed on a 100MW biomass fired electricity generating boiler at the Gainesville Renewable Energy Center.

The combustion of fossil fuels is the leading contributor to global air emissions. A variety of APC technologies are used to limit these stack emissions, as required by the Clean Air Act and other legislation. Almost without exception, these technologies rely upon appropriately optimized fluid dynamic performance to ensure the required process performance. A common example of this reliance upon fluid dynamics optimization is that of SCR systems. SCR systems are dependent upon the effective homogenization of an injected reagent within the NOx containing flue gas to achieve high NOx reduction efficiency across the catalyst bed. The authors' development of numerical tools to evaluate the impact of non-homogeneous velocity and concentration profiles on NOx reduction performance have enabled a substantially more accurate assessment of the importance of this optimization and will be presented for the first time here. In addition, the specific combination of technologies that enable the effective homogenization including reagent injection equipment, static mixers and other flow distribution devices, will be discussed in light of the performance enhancements achieved. Of recent interest in this area of focus has been the coupling of CFD simulations to evolutionary search algorithms, such as genetic algorithms. The authors' progress in the application of this methodology will be discussed with an example given demonstrating Pareto front convergence in a constrained multi-objective search.

421520 A CFD-DEM Model for Viscous Solid-Liquid Mixing in Stirred Tanks

Abstract:

Viscous solid-liquid mixing plays a key role in the production of a large variety of consumer goods such as pastes, paints, cosmetics, pharmaceuticals and food products. Despite this industrial relevance, the majority of reported results in solid-liquid mixing has been geared towards the turbulent regime, and little is known about the laminar and transitional regimes. In particular, it remains unclear how the rheology of a suspension, the particle interactions and the presence of a complex rotating geometry impact the flow patterns as well as critical parameters such as the impeller torque, the just-suspension speed (Njs) and mixing indices. To shed light on these issues, numerical and experimental work is essential.

A variety of models have been developed to simulate solid-liquid flows. These include the Eulerian-Eulerian model, and the combination of the Discrete element method (DEM) for the particles and CFD methods for the liquid phase (CFD-DEM). While it possesses a huge potential due to its formulation, notably as regards its natural capacity to reproduce the maximal packing fraction of solid particles, the ability of the CFD-DEM approach to accurately model viscous solid-liquid flow in complex geometries has yet to be assessed, and the method has not been validated experimentally in the field of mixing.

In the present work, we extend the so-called CFDEM framework, which combines OpenFOAM and LIGGGHTS, to study viscous solid-liquid mixing. First, the governing equations for the liquid and the solid phases are presented along with a two-way coupling strategy. The model is next applied to the study of solid-liquid mixing in a stirred tank equipped with a pitched-blade turbine (PBT). In particular, experimental visualization of flow patterns is used to assess the accuracy of the proposed model. Finally, the model is validated quantitatively by comparing the fraction of suspended solid measured as a function of the impeller rotational speed.

423864 Investigating Bubble Size Distribution in Fermentation Reactors with CFD

Abstract:

Fermentation is a common method for producing bio-products such as alcohols and pharmaceutical compounds or alternative feedstocks. In the primary stage, presence of oxygen drives the metabolism and hence growth of the organic substance. Oxygen is supplied via air in stirred vessel type fermenters. While a certain configuration tested in a lab may result in satisfactory distribution of the dissolved oxygen (DO), large scale fermenters need careful design considerations.

Computational methods (CFD) have been reported in large amount of literature to investigate the flow patterns and consequently the gas hold up in such equipment. However, the air bubble injected vary in size throughout the reactors due to the choice of impellers, rpm, baffles, and other internals such as coils or dip tubes. The size of the air bubbles affects directly the gas hold up and the interfacial area. Therefore it is necessary to account for the varying bubble size distribution due to coalescence and breakup.

In this work, we present the use of a size distribution model to account for the bubble size distribution called S-gamma model in the commercial code STAR-CCM+. This model shows the differences in assuming constant bubble size but is not as computationally expensive as the quadrature (DQMOM/QMOM) type of models.

Features of the S-gamma model will be discussed and a case study presented where the two methods are compared. An outlook on future work and advances will also be given.

References:

Lo, S. and Zang, D., Modelling of Break-up and Coalescence in Bubbly Two-Phase Flows, Journal of Computational Multiphase Flows, Vol 1(1), p 23, 2009

430288 CFD Simulations of a Stirred Tank Reactor for Liquid and Gas-Liquid Systems Using Openfoam®

Abstract:

Stirred tank reactors are widely used in industrial processes, including chemical processes like polymerization, biological processes such as cell culture and fermentation, pharmaceutical industry, and many other processes. Multiple impellers are usually preferred over single impeller for tall-thin configuration of industrial reactors for better gas utilization and maintenance of homogeneity in the reactor. In this work, the stirred tank reactor experimentally studied in Shewale and Pandit [1] is investigated numerically. The reactor is constituted by a cylindrical vessel with four baffles, and three impellers (one Rushton turbine, at the bottom of the tank, and two pitched blade turbines) mounted on the shaft and rotating synchronously.

A three-dimensional hexahedral mesh for the system is generated using a commercial mesh-generation software Pointwise®. CFD simulations are performed with the open-source CFD software OpenFOAM®. Steady state single-phase (liquid) simulations are performed with the realizable k-ε model, in order to describe the turbulence in the reactor. For gas-liquid system, a two-fluid model [2] with the mixture k-ε turbulence model [3] is used to solve the flow. The multiple reference frame approach [4] is applied for both systems to treat the rotation of impellers in the reactor. For both single phase and multiphase systems, the liquid phase is water at 300 K. And for multiphase system, air at 300 K is injected uniformly from the ring sparger located below the bottom Rushton turbine. Simulation results are compared to experiments reported in Shewale and Pandit [1], including flow patterns, unaerated and aerated power consumption, and overall gas holdup. The effects of different operating conditions such as impeller rotation speed at 3.75 rad/s and 5.08 rad/s are also studied.

[1] S.D. Shewale, A.B. Pandit, Studies in multiple impeller agitated gas–liquid contactors, Chemical Engineering Science. 61 (2006) 489–504.

[2] D.A. Drew, Mathematical modeling of two-phase flow, Annual Review of Fluid Mechanics. 15 (1983) 261–291.

[3] A. Behzadi, R.I. Issa, H. Rusche, Modelling of dispersed bubble and droplet flow at high phase fractions, Chemical Engineering Science. 59 (2004) 759–770.

[4] J. Luo, R. Issa, A. Gosman, Prediction of impeller induced flows in mixing vessels using multiple frames of reference, in: Institution of Chemical Engineers Symposium Series. 136 (1994) 549.

431353 Validation of Multiphase Mean Age Theory

Abstract:

A new method for determining mean age in multiphase mixtures has been developed and validated against experimental data. While preliminary results have been presented previously connecting theory to application, this study focuses on experimental validation of the fundamental theory to provide a rigorous foundation upon which to build future applications.

Scalar transport equations were solved to track a secondary phase independently from the bulk fluid. Each phase has its own scalar associated with it. In this way, mean age and residence time distributions (RTDs) can be determined for each phase separately. A two-phase oil/water system flowing in a pipeline was simulated to determine the RTDs at multiple locations for both phases using our steady state multi-phase mean age technique. These data were compared against the RTDs generated from a traditional transient CFD simulation, and verified against experimental results. All three methods generated RTDs with similar mean residence times and variances, with differences as low as 4%.

This method provides the same benefits of traditional mean age theory, primarily obtaining local time-dependent data from steady state simulations. Many important long time-scale applications are envisioned such as platelet settling in the human vasculature and diseased hearts, solids settling in lakes and streams, pollutant dispersion in air, and solids dispersion in chemical reactor systems.

431715 Numerical Investigation of Bubble Effects on Taylor-Couette Flow Patterns in the Weakly Turbulent Vortex Flow Regime

Abstract:

Bubbly gas-liquid Taylor-Couette vortex flow has been the subject of several recent investigations both because of interest in bubble-induced drag reduction and because such devices have potential applications to a variety of chemical and biochemical processing problems. The effect of bubbles on Taylor-Couette flow patterns and structures in weakly turbulent vortex flow regime is investigated by CFD simulation. The CFD model is based on an Eulerian-Eulerian two-fluid approach with constitutive closures for inter-phase forces and liquid turbulence and validated by comparing simulation predictions with previously published experimental data. The bubble spatial distribution patterns and mechanism by which the dispersed fluid reaches a non-uniform radial distribution is discussed, and the relative importance of various inter-phase forces is demonstrated. In addition, the axial expansion of Taylor vortices due to the presence of bubbles was reproduced by the simulations, and the influence of azimuthal Reynolds number and gas flow rate on vortex axial wavelength, liquid velocity and turbulence are discussed in detail.

436211 Characterization of Gas Dispersion Behavior in Flotation Cells Using Coupled CFD-PBM and Direct PBM Approaches

Abstract:

Accurate predictions of Sauter mean diameter (SMD) of dispersed phase (gas) is necessary to facilitate accurate predictions of flotation kinetics in the flotation cells. Previous studies have shown the presence of spatial non-uniformity in bubble size distribution (BSD) for both lab- and pilot-scale flotation cells. In this study, we aim to use coupled computational fluid dynamics (CFD) and population balance modeling (PBM) approach to predict the distribution of SMD in generic flotation cells. The dynamics of multiphase flow is modeled using Eulerian-Eulerian approach and BSD is obtained by using quadrature method of moments (QMOM). Only binary breakage and coalescence processes of gas bubbles are considered. Coalescence rate is modeled using Coulaloglou and Tavlarides model (1977), while breakage rate and daughter size distribution are modeled using Laakkonen (2007) model. A moment correction procedure proposed by McGraw (2012) is used to identify and correct invalid moment sequences in each computational cell. The numerical approach applied in this work is validated by comparing the numerical predictions against published BSD in stirred tanks. Using the coupled CFD-PBM approach, overall gas holdup and SMD for flotation cell are obtained. Furthermore, flotation cell is also modeled as a homogeneous well-mixed reactor, and the resulting population balance equation (PBE) is solved using high-order moment conserving method of classes (HMMC) technique. Different breakage and coalescence models are tested, and their effect on final BSD and SMD are investigated. Finally, the predictions obtained from coupled CFD-PBM and HMMC techniques are compared for the entire cell.

431956 CFD Investigation of Gas-Liquid Mass Transfer in a Multiphase Taylor-Couette Reactor

Abstract:

Bubbly gas-liquid Taylor-Couette vortex flow has gathered increasing interest because of its potential applications to a variety of chemical and biochemical processing problems. In order to quantitatively describe the hydrodynamics and mass transport during two phase Taylor-Couette flow, a rigorous two-fluid CFD incorporating interphase mass transfer was developed for the simulation of bubbly Taylor-Couette flow and compared with data obtained from oxygen transport experiments using an air-water Taylor vortex flow system. Important parameters influencing the mass transfer coefficient were analyzed, including bubble Sauter mean diameter, slip velocity and turbulent dissipation rate. Several mass transfer models, such as the penetration model and eddy cell model, were compared tested against experimental data. Subsequently, an adaptive simulation method was developed that automatically selects the most appropriate mass transfer model based upon local flow conditions, leading to more accurate predictions for the global rate of mass transfer. Predictions of this adaptive method are in good agreement with experimental findings for a wide range of gas flow rates and azimuthal Reynolds numbers in gas-liquid Taylor-Couette flow.

437406 Correlation of Non-Baffled, Angled and Off-Center Agitated Process Vessel Blend Times

Abstract:

Non-baffled vessels are appealing for appropriate light duty mixing applications in the pharmaceutical industry because they may simplify the equipment cleaning challenges associated with baffles. The use of angled and off-center axial flow mixers in non-baffled dished bottom vessels can provide light duty mixing with good vertical mixing and minimal swirling. The blend time to achieve 95% homogeneity for a family of geometrically similar tanks from 100L to 5,000L from a manufacturing facility was characterized and fitted to multiple published engineering literature correlations which have been previously developed for baffled tanks. A comparison of the data fits to the correlations will be presented.

425170 Experimental Determination of Solids Suspension with Angled Impellers in Pharmaceutical Mixing Vessels

Abstract:

Angled-mounted impellers are commonly used in a number of pharmaceutical industry applications, from laboratory reactors to full-scale tanks. In these systems, the impeller is not centrally and vertically mounted, as in most stirred vessels, but it enters the vessel, and is therefore immersed in the fluid, at an angle from the vertical. This arrangement provides some baffling effects that would not otherwise exist if the impeller was centrally mounted.

One common requirement for such systems is that they are capable of suspending small solids dispersed in the liquid. Therefore, it is critical to know the value of the minimum agitation speed at which solids become just suspended, N_{js}. Despite their common use in industry, and especially the pharmaceutical industry, these systems have received little attention in the literature in general and as far as their ability to solid suspension is concerned in particular.

Therefore, the focus of this work was to determine under what conditions fine divided solids become suspended in a liquid-filled, unbaffled stirred vessel provided with angled-mounted impellers. A small laboratory vessel with a hemispherical bottom was used here. Different types of axial and radial impellers mounted at different angles (0°, 5° and 10°) at different off-bottom impeller clearances were tested. The value of N_{js} was experimentally obtained by visually inspecting the tank bottom and determining the impeller agitation speed at which the solids were observed to rest on the tank bottom for no more than 1-2 seconds before being swept away. In addition, a novel observer-independent criterion was developed, relying on the experimental determination of the fraction of solids remaining on the vessel bottom at increasing agitation speeds. This approach results in values of N_{js} that are close agreement with the visually determined N_{js}values.
This approach was used to determine N_{js} for different impellers under a number of operating conditions. In general, mounting the impeller at an angle resulted in N_{js} values lower than those obtained with centrally mounted impellers. In addition, it was found that lower impeller clearances resulted in lower N_{js} values, although the minimum for $N_{js}$was obtained when the impeller clearance was about 20% of the vessel diameter.

The results obtained in this work are expected to be applicable to a number of systems currently used in the pharmaceutical industry.

427057 Multiphase Bioreactor Simulation with GPUs

Abstract:

The effectiveness of a biopharmaceutical manufacturing process depends to a large extent on the efficiency of the bioreactor, especially in the field of generic drugs. The improvement of the productivity of industrial size bioreactors is, up to now, mainly driven by empirical knowledge. Computer simulation would help to understand the processes inside the reactor and hence lay the foundation for new reactor designs. However, despite the recent improvement of computational capabilities, only a few seconds of real time operation of an industrial scale reactor takes months to simulate.

Thus, the goal of this study is to use graphic cards to speed up this simulation. To take full advantage of the processor's parallelism, localized calculation algorithms are used, like the lattice Boltzmann method for modelling the fluid flow field. The calculation of the bubble movement is done by solving the Newton's equation of motion. For coalescence and breakup stochastic models are used. The phases are coupled with a two way approach.

To simulate large reactors, the code has a multi GPU functionality, hence it can distribute the workload of the simulation on several graphic processors. Validation is done with literature data as well as with results from holdup measurements of a 150l laboratory reactor. To show the usefulness of this code for scale up studies, a 40 m^3 reactor was simulated.

To mimic the behavior of real fermentation broth a non-Newtonian viscosity model was added using the power law model. An important process within the reactor is the transport of species (dissolved oxygen, nutrients, carbon dioxide, etc.) by the fluid flow field. To model the distribution of the species a particle based solver was implemented.

434243 Retreat Curve Impeller Flow and Heat Transfer Characteristics

Abstract:

Glass-lined vessels and impellers are often encountered in mixing operations, particularly in the pharmaceutical and fine chemicals industries. Retreat Curve Impellers are designed without sharp edges to make them easy to coat with glass. Although other more efficient glass-coated impeller designs are currently available, the retreat curve impeller is still in common use. A number of studies, both experimental and computational, have been conducted to characterize the flow patterns, mixing times, power requirements and just suspended speeds of mixing systems using these impellers. Very few studies have included the heat transfer characteristics. This talk will present the procedures and results of a previously unpublished experimental investigation to develop tables and correlations for the Power Number, Flow Number, and Nusselt Number associated with a dished bottom vessel fitted with a retreat curve impeller of various diameters and off-bottom clearances, as well as different baffle arrangements. The heat transfer data was obtained using a novel wall heat flux system with data for the vessel sides and bottom treated separately. These correlations have been used to design and optimize numerous stirred tank vessels and processes.

427476 Advances in Gas Sparging Simulation for Bioreactor Modeling to Create Comprehensive "Simulation Design Space"

Abstract:

Gas-liquid dispersion in stirred tanks is a commonly used industrial process, e.g., bio-pharmaceutical or fine-chemical manufacturing. While CFD models have been used to predict gas holdup, bubble size distribution and mass transfer coefficients, there have been fewer studies on deploying the CFD models to range of operating conditions. This study focuses CFD simulation at different rotational speed (rpm) and gas flow rates to create a comprehensive "Simulation Design Space" for mass transfer co-efficient (Kla). Mass transfer co-efficient is experimentally obtained by measuring Dissolved Oxygen (DO) concentration by absorption in pure water. In current study, simulation model is developed numerically to calculate DO through multiphase and multispecies CFD simulations. Population balance model is coupled with the Eulerian multiphase-multispecies model to predict the bubble size distribution and oxygen distribution. The study also looks into effects of turbulence on drag co-efficient which is a key factor in predicting an accurate bubble size distribution and gas hold-up. Simulation best practices are developed and then deployed for validation of the model against experimental data [1] on wide range of operating conditions.

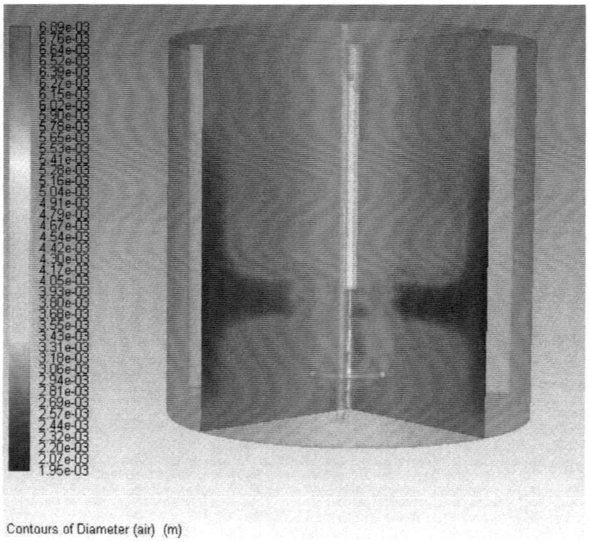

Bubble size distribution on planes between baffles

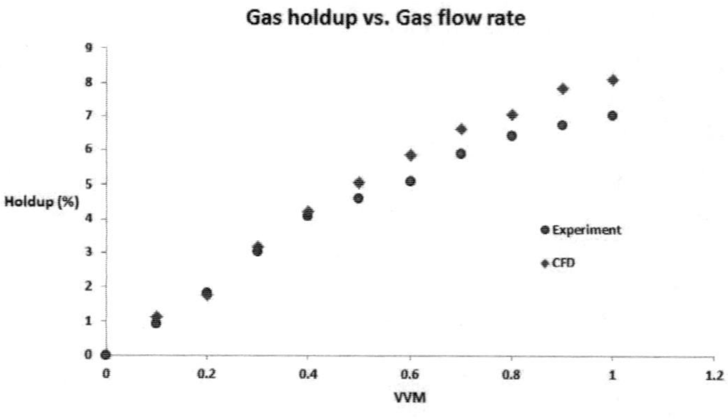

Variation of gas holdup with gas flow rate

1. M. Laakkonen et al., Chemical Engineering Science 62 (2007) 721 – 740

425747 Particle Engineering of Cocrystals Using a Solvent-Free Approach By Spray Congealing

Wednesday, November 11, 2015: 12:30 PM
Ballroom D (Salt Palace Convention Center)

Íris Duarte[1,2], João F. Pinto[1] and **Márcio Temtem**[2], (1)iMed.ULisboa, Faculty of Pharmacy, University of Lisbon, Lisbon, Portugal, (2)R&D Drug Product Development, Hovione FarmaCiência SA, Loures, Portugal

Particle engineering of cocrystals using a solvent-free approach by spray congealing

Iris Duarte[1,2], Joao F. Pinto[2], Marcio Temtem[1*]

[1] Hovione Farmaciencia SA, Sete Casas, 2674-506 Loures, Portugal;
[*]mtemtem@hovione.com or +351 219 847 569

[2] iMed.ULisboa, Faculdade de Farmacia da Universidade de Lisboa, Av. Prof. Gama Pinto, 1649-003, Lisboa, Portugal

Around 90% of the new chemical entities emerging from the drug discovery phase present, at least, solubility constraints (BCS Class II+IV). Among the different solubilization strategies that have been applied to address this issue, the use of pharmaceutical cocrystals emerged as an alternative crystal-engineering platform to enhance the poor physicochemical properties of these challenging molecules [1].

Pharmaceutical cocrystals are stable multicomponent crystals, typically comprising two molecules, *e.g.* an API, or the salt of an API, and a coformer. The coformer can be an active or non-active ingredient (*e.g.* pharmaceutical excipients, vitamins, minerals, amino acids, other APIs) that has a favorable intermolecular interaction with the API, promoting hydrogen bonding, van der Wall forces or pi-stacking [2].

In general, and despite the great potential of cocrystallization, producing cocrystals with high efficiency at industrial scales is challenging due to the tight thermodynamic and kinetic crystallization windows necessary to achieve the desired crystalline form. Specifically referring to the pharmaceutical field, the wide application of cocrystals is still limited due, in part, to the scarcity of suitable large-scale production methods and lack of robust and cost-effective processes.

In order to address some of these challenges, spray congealing has recently emerged as new process for the preparation of cocrystals [3]. Spray congealing, which is a hybrid technology governed by similar principles of spray drying and hot melt extrusion, consists in feeding a molten mixture to an atomizer, which then breaks the liquid feed into small droplets, and those droplets are cooled thanks to a co-current stream of cooling gas (Figure 1).

The advantages of this method are the fact it can be conducted in the same apparatus of spray drying, with minor modifications, thus making the commercial scale of the process readily available; for being a solvent free technique, cocrystallization via spray congealing will also comply with green chemistry requirements and, most importantly, it can be used to produce "engineered" cocrystals (*i.e.* with the desired properties) in a particulate form, in a single stage operation without the need of any downstream processing.

The feasibility of using spray congealing to produce pharmaceutical cocrystals was successfully demonstrated using three model API:coformer systems (Table 1).

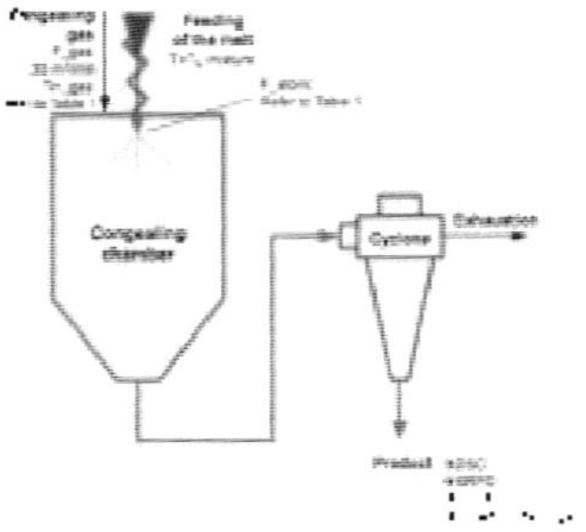

Figure 1. Schematic representation of the spray congealing set up used.

Table 1. Case-studies and respective process variables defined for the tests. The flow rate of the congealing (F_gas) was set at $0.35 m^3/min$.

System	Molar Proportion	Process Variables		Ref
		ΔT [°C]*	F_atom [L/min]**	
Caffeine:Glutaric acid (CAF:GLU)	1:1	0-50	11-20	4,5
Caffeine:Salicylic acid (CAF:SAL)	1:1	50	9	5
Carbamazepine:Nicotinamide (CBZ:NIC)	1:1	125	12	5-7

* ΔT: difference between the mixture melting temperature observed inside the beaker ($T_{M,mix}$) and the inlet temperature of the congealing gas ($T_{in,\ gas}$);

**F_atom: flow of the atomization gas.

All the final products obtained with spray congealing presented DSC thermal profiles and XRPD diffractograms characteristic of the respective cocrystals and different from the pure components or physical mixtures. As an example, Figure 2 shows the results obtained for the 1:1 Caffeine: Salicylic Acid system.

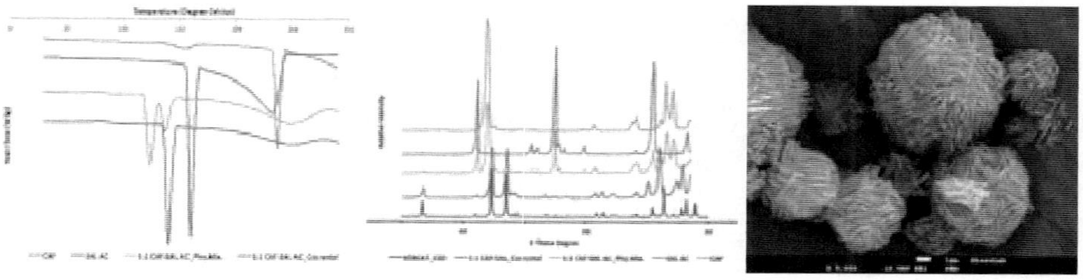

Figure 2. Results obtained for the 1:1 CAF:SAL system. From left to right: DSC, XRPD and SEM.

Moreover, and for one specific case-study, *i.e.* 1:1 Caffeine:Glutaric acid (CAF:GLU), a DoE with 2 parameters at 2 levels + 1 central point was conducted, to assess the effect of process variables on the final cocrystal particle and bulk powder properties. The parameters or process variables (*i.e.* ΔT and F_atom) were varied according to the ranges shown in Table 1.

According to Figure 3 it can be observed that particle properties (size, morphology, shape) can be adjusted, *in situ*, as part of the spray congealing process, without compromising cocrystal formation (DSC and XRPD data not shown).

Moreover, particle properties are known to affect the bulk behaviour of powders, which can impact subsequent downstream processing (*i.e.* blending, capsule filling, tableting, etc.) [8]. In order to evaluate this, bulk (*i.e*, compressibility and permeability) and shear properties of the powders produced were measured using a FT4 powder rheometer. Figure 4 shows the results from the permeability and compressibility tests at 15 kPa, and the shear cell test on samples that have been pre-consolidated at 9kPa normal stress.

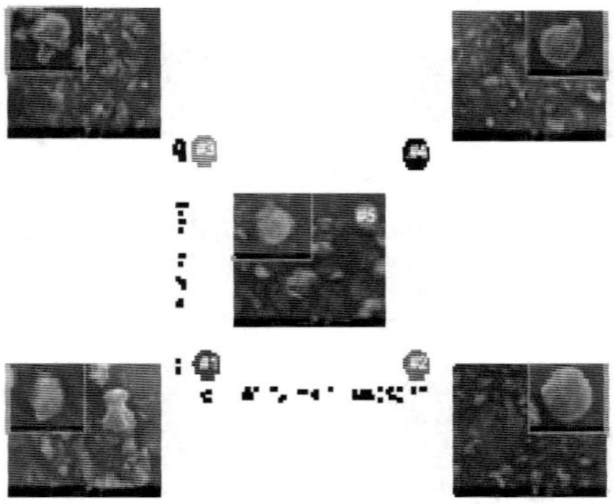

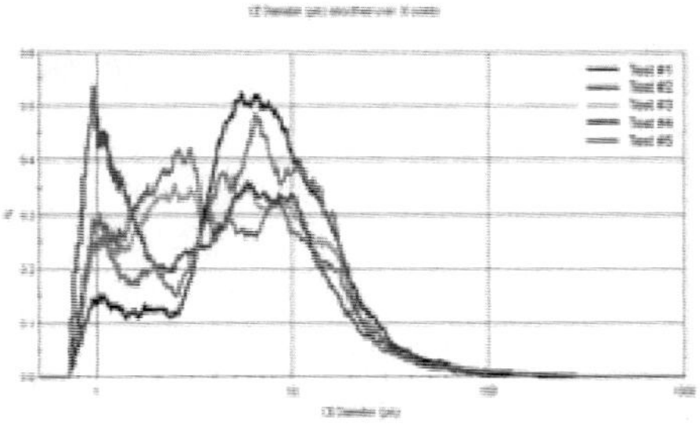

Figure 3. SEM results (right) and number-based circular equivalente diameter (CED) distributions (above) for the 1:1 CAF:GLU systems produced, when varying ΔT and F_atom.

Shear testing classified the powders as identical, whilst the permeability and compressibility tests show they are significantly different. The shear data suggest that the powders may behave similarly in a moderate to high stress environment, such as in a hopper, but it does not necessarily mean that they will behave similarly when aerated or flowing under low stress. The different tendencies of the permeability and compressibility results are clearly a consequence of the influence of multiple variables or particle properties (*e.g.* particle size, shape, surface properties, amount of fines, etc.) on the bulk powder rheology.

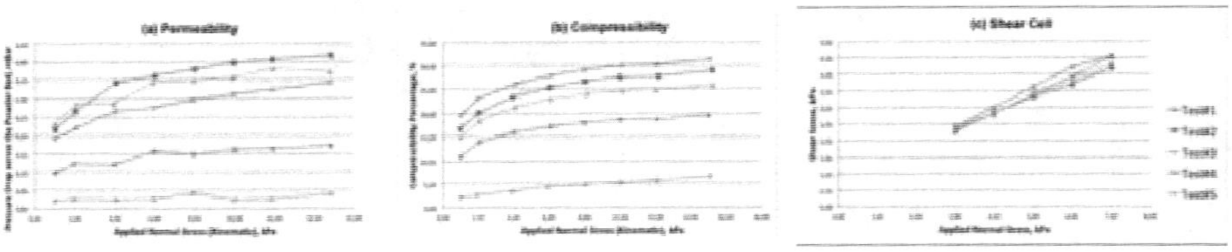

Figure 4. From left to right: permeability as a function of normal stress at constant air velocity of 2 mm/s, compressibility (*i.e.* % of bulk density change) as a function of normal stress and shear stress as a function of applied normal stress.

References:

[1] Blagden N. *et al.*, Pharmaceutical co-crystals - are we there yet?. *CrystEngComm*, 2014, 16, pp. 5753-5761;

[2] Qiao, N. *et al.*, Pharmaceutical cocrystals: An overview. *International Journal of Pharmaceutics*, 2011, 419, pp. 1-11;

[3] Duarte, I., *et al.* Synthesis and particle engineering of cocrystals. WO2015/036799A1 (2015);

[4] Trask *et al.*, Solvent-drop grinding: green polymorph control of cocrystallization. *Chem. Commun.*, 2004, 7, 890-891;

[5] Lu *et al.*, A rapid thermal method for cocrystal screening. *CrystEngComm*, 2008, 10, 665-668;

[6] Chieng *et al.*, Formation Kinetics and Stability of Carbamazepine - Nicotinamide Cocrystals Prepared by Mechanical Activation. *Cryst. Growth Des.,* 2009, 9(5), 2377-2386;

[7] Liu *et al.*, Improving the Chemical Stability of Amorphous Solid Dispersion with Cocrystal Technique by Hot Melt Extrusion. *Pharm. Res.*, 2012, 29, 806-817;

[8] Freeman R., Measuring the flow properties of consolidated, conditioned and aerated powders – a comparative study using a powder rheometer and rotational shear cell. *Powder Technology*, 2007, 174, pp. 25-33.

433246 A Case Study in Tuning API Powder Properties Via Crystallization to Meet Formulation Needs

Abstract:

Crystal form and powder properties are universal quality attributes which characterize every API. This presentation is a case study of one API that has the potential to form numerous polymorphs and a variety of crystal morphologies. Process development efforts were focused on obtaining the thermodynamically stable crystal form while targeting a narrow range of powder properties suitable for bioavailability and downstream drug product formulation. The work includes screening studies to select the appropriate solvent system and crystallization parameter optimization to control particle size and minimize agglomeration. Multiple technologies were employed to support this work, including FTIR, FBRM, wet milling, and dry milling. Understanding the impacts of crystallization parameters allowed for the generation of several lots of material with a variety of powder properties, which were evaluated for formulation behavior and bioavailability. The feedback and collaboration between drug substance and drug product process development teams enabled the selection of the desired target material properties (eg. particle size distribution, D90/D50 ratio) and formulation process (roller compaction or wet granulation) that would robustly lead to high quality drug to patients.

432520 Improving Dissolution Kinetics of Pharmaceuticals By Fluidized Bed Impregnation

Abstract:

According to the current Biopharmaceutics Classification System (BCS) adopted by the U.S. Federal Drug Administration (FDA), drug substances are categorized into four main classes, based on their solubility and permeability. Two of those classes (BCS classes II and IV) are characterized with low solubility. The trend in the past decade has been toward steady increase in the number of active pharmaceutical ingredients (APIs) belonging to BCS classes II and IV. It has been estimated that up to 60% of new chemical entities fall in the low solubility class, compared to 39% for marked drugs. This trend towards decrease in API solubility is undoubtedly one of the main hurdles for formulators today, as they continuously strive to develop novel formulations capable of effectively delivering the drug into the blood stream.

API solubility has a direct effect on the drug's bioavailability and therefore its efficacy. The absorption of BCS Class II and IV compounds and the resulting blood concentration-time profile depends on the rate-limiting step, which can be one of the following: permeability limited, dissolution limited and solubility limited. For poorly-soluble compounds, improving the rate (for dissolution limited) and the extent of dissolution (for solubility limited) will directly increase their bioavailability. There are two main techniques that are widely used today for improving dissolution rate and increasing solubility. The first technique involves particle size reduction as suggested by the Nernst-Brunner/Noyes-Whitney equation, where the dissolution rate for any API is proportional to the surface area available for dissolution. Solubility of any substance is a thermodynamic property, which among other environmental factors (solvent nature, pH and temperature) depends strongly on the chemical or physical structure of the substance itself (e.g. salts, crystalline or amorphous solids, solvates, co-crystals). Many variations of these two main techniques for improving dissolution rate and increasing solubility are employed into today's development or marketed drug formulations. The most common include various bottom up or top down approaches for size reduction and numerous means for producing amorphous materials (e.g. spray drying, freeze drying, hot-melt extrusion, polymer additives). Often these methods have major challenges and limitations associated with: difficulty in control or scale up, API stability, applicability to wide range of APIs, high cost, etc.

In this work we present a drug formulation method that is capable of increasing significantly the dissolution kinetics of a poorly soluble API (Fenofibrate), which shows a high degree of simplicity compared to many of the traditional methods. The method involves impregnation of the API into highly porous excipient in a fluidized bed (FB). Loadings of up to 40% were successfully formulated, showing significant improvement in the dissolution profile. Additional improvement in dissolution was achieved by mild milling of the impregnated drug product. Moreover, co-impregnation of the API with an FDA approved additive showed significant further improvement in dissolution profile, making our formulation comparable to one of Fenofibrate's patented market formulations (TriCor, 45mg). Other benefits of our impregnation method include: high blend uniformity independent on API nature or loading, excellent flow properties of the impregnated powder (no need for more additives), elimination of several unit operations from drug substance and drug product development.

423877 A Drop-on-Demand Manufacturing System for the Production of Amorphous Drug Products

Abstract:

Innovation is the key to success within the pharmaceutical industry. However, it has been largely sought through new drug discovery and development, while the development of more efficient manufacturing methods has received inadequate attention. Recently, as a result of encouragement from the regulatory authorities, the advent of globalization and the increasing awareness of environmental impact, the pharmaceutical industry has been reconsidering the way drug products are developed and manufactured. The US Food and Drug Administration (FDA) has promoted the Quality by Design (QbD) approach to increase process understanding and improve quality and efficiency while minimizing risk [1]. The Process Analytical Technology (PAT) guidance introduced by the FDA encouraged the monitoring of critical quality and performance attributes during processing with the goal of ensuring final product quality [2,3].

As part of this renewed emphasis on improvement of manufacturing and under the US National Science Foundation supported Engineering Research Center for Structured Organic Particulate Systems; we have developed a dropwise additive manufacturing process for solid oral dosage production. The process utilizes the drop-on-demand (DoD) inkjet printing technology for predictable and highly controllable deposition of active pharmaceutical ingredients (API) onto an edible substrate, such as a polymeric film or placebo tablet [4, 5]. This process uses fluid operations suitable for low volume production of personalized dosage forms. The advantages of liquid processing and reproducible production of small droplets create the special opportunity for the production of high potency, low dose drugs, which are difficult to produce with consistent quality via conventional powder processing methods.

Using the developed drop on demand manufacturing process for pharmaceuticals, different drug formulations including solvent based systems [4], i.e. solvent-polymer-API solutions, as well as melt based systems [5], i.e. polymer-API melts, can be printed. Melt-based printing applications eliminate the solvent evaporation step after drop deposition and thus allow on demand production of individual dosage forms with good control of drug solid state and morphology.

In this paper, we present the controlled production of amorphous dosages by the deposition of melt-based formulation using the DoD manufacturing system. Producing dosage forms in which the active ingredient is in amorphous forms can enhance the dissolution of low-solubility drugs. Amorphous drug forms are produced with alternative choice of polymers, which inhibit crystallization, along with suitable choice of operating conditions. Polymers are added to the printing material to help control drug morphology, material properties and the formulation composition [6]. Addition of crystallization inhibiting polymers, such as PVP, to the formulation, can result in the formation of dosage forms containing amorphous or semi-crystalline API. Operating conditions, i.e. crystallization temperature, are controlled using substrate temperature controller, which is used to achieve the desired product quality by influencing the dissolution behavior and therefore the bioavailability of the drug [7]. Since stability of amorphous drug forms can be challenging, design of crystallization temperature profile and precise control of substrate temperature profile are even more important than in the case of crystalline API formulations. The amorphous dosage forms are achieved by controlling the cooling profile of the deposits to solidify the drug in the amorphous form before crystallization occurs.

1. Yu LX, Amidon G, Khan MA, Hoag SW, Polli J, Raju GK, et al. Understanding pharmaceutical quality by design. AAPS J. 2014;16(4):771–83.
2. Food and Drug Administration CDER. Guidance for Industry PAT - A Framework for Innovative Pharmaceutical. 2004.
3. Simon LL, Pataki H, Marosi G, Meemken F, Hungerbu K, Baiker A, et al. Assessment of Recent Process Analytical Technology (PAT) Trends: A Multiauthor Review. Org Process Res Dev. 2015;19(1):3–62.
4. Hirshfield L, Giridhar A, Taylor LS, Harris MT, Reklaitis G V. Dropwise additive manufacturing of pharmaceutical products for solvent-based dosage forms. J Pharm Sci. 2014;103(2):496–506.
5. Içten E, Giridhar A, Taylor LS, Nagy ZK, Reklaitis G V. Dropwise Additive Manufacturing of Pharmaceutical Products for Melt-Based Dosage Forms. J Pharm Sci. 2015;104(5):1641-49.
6. Trasi NS and Taylor LS. Effect of Additives on Crystal Growth and Nucleation of Amorphous Flutamide. Crystal Growth & Design 2015; 12(6): 3221–3230.
7. Içten E, Nagy ZK, Reklaitis GV. Supervisory Control of a Drop on Demand Mini-manufacturing System for Pharmaceuticals. Comp. Aided Process Eng., 2014; 33: 535-540.

441851 Introduction of Advances in Industrial Mixing

Abstract:

Advances in Industrial Mixing is a companion volume and update to the Handbook of Industrial Mixing. The second volume fills in gaps for a number of industries that were not covered in the first edition. Significant changes in five of the fundamental areas are covered in entirely updated or new chapters. The original text is provided as a searchable pdf file on the accompanying USB. * This book explains industrial mixers and mixing problems clearly and concisely. * Gives practical insights by the top professionals in the field, combining industrial design standards with fundamental insight. * Details applications in 14 key industries. Six of these are new since the first edition. * Provides the professional with information he/she did not receive in school. * Five completely rewritten chapters on mixing fundamentals where significant advances have happened since the first edition and seven concise update chapters which summarize critical technical information.

441852 Improved Correlation for Predicting Just Suspension Speed in Stirred Tanks

Abstract:

Advances in Industrial Mixing is a companion volume and update to the Handbook of Industrial Mixing. The second volume fills in gaps for a number of industries that were not covered in the first edition. Significant changes in five of the fundamental areas are covered in entirely updated or new chapters. The original text is provided as a searchable pdf file on the accompanying USB. * This book explains industrial mixers and mixing problems clearly and concisely. * Gives practical insights by the top professionals in the field, combining industrial design standards with fundamental insight. * Details applications in 14 key industries. Six of these are new since the first edition. * Provides the professional with information he/she did not receive in school. * Five completely rewritten chapters on mixing fundamentals where significant advances have happened since the first edition and seven concise update chapters which summarize critical technical information.

441853 Mixing Processes and Process Validation in the Pharmaceutical Industry

Abstract:

The *Advances Volume* is a companion to the *Handbook of Industrial Mixing.* It reworks areas that have seen significant progress since the first edition and includes:

- Brand new chapters for 6 areas of industrial application: pharmaceutical validation, crystallization, water treatment, commissioning of mixing equipment, and safety
- Additional sections or examples for 5 chapters from the first edition including: magnetic drives, micromixers, vessel head volume calculations, additional information on heat transfer, and application of the Bourne protocol to reactor design
- Concise but critical technical updates for mixing in pipelines, turbulence, and CFD
- 5 completely new fundamentals chapters covering: mean age distrbution, rheology and complex fluids, flow patterns, viscous mixing, solid-liquid mixing
- A full technical definition of mixing
-

441855 Heat Transfer in Stirred Tanks - Update

Abstract:

Advances in Industrial Mixing is a companion volume and update to the Handbook of Industrial Mixing. The second volume fills in gaps for a number of industries that were not covered in the first edition. Significant changes in five of the fundamental areas are covered in entirely updated or new chapters. The original text is provided as a searchable pdf file on the accompanying USB. * This book explains industrial mixers and mixing problems clearly and concisely. * Gives practical insights by the top professionals in the field, combining industrial design standards with fundamental insight. * Details applications in 14 key industries. Six of these are new since the first edition. * Provides the professional with information he/she did not receive in school. * Five completely rewritten chapters on mixing fundamentals where significant advances have happened since the first edition and seven concise update chapters which summarize critical technical information.

441856 Microstructure, Rheology, and Processing of Complex Fluids

Abstract:

Advances in Industrial Mixing is a companion volume and update to the Handbook of Industrial Mixing. The second volume fills in gaps for a number of industries that were not covered in the first edition. Significant changes in five of the fundamental areas are covered in entirely updated or new chapters. The original text is provided as a searchable pdf file on the accompanying USB. * This book explains industrial mixers and mixing problems clearly and concisely. * Gives practical insights by the top professionals in the field, combining industrial design standards with fundamental insight. * Details applications in 14 key industries. Six of these are new since the first edition. * Provides the professional with information he/she did not receive in school. * Five completely rewritten chapters on mixing fundamentals where significant advances have happened since the first edition and seven concise update chapters which summarize critical technical information.

441857 Flow Patterns and Mixing - Visualization

Abstract:

Advances in Industrial Mixing is a companion volume and update to the Handbook of Industrial Mixing. The second volume fills in gaps for a number of industries that were not covered in the first edition. Significant changes in five of the fundamental areas are covered in entirely updated or new chapters. The original text is provided as a searchable pdf file on the accompanying USB. * This book explains industrial mixers and mixing problems clearly and concisely. * Gives practical insights by the top professionals in the field, combining industrial design standards with fundamental insight. * Details applications in 14 key industries. Six of these are new since the first edition. * Provides the professional with information he/she did not receive in school. * Five completely rewritten chapters on mixing fundamentals where significant advances have happened since the first edition and seven concise update chapters which summarize critical technical information.

441858 Advances in Industrial Mixing: We've Come a Long Way!

Abstract:

Advances in Industrial Mixing is a companion volume and update to the Handbook of Industrial Mixing. The second volume fills in gaps for a number of industries that were not covered in the first edition. Significant changes in five of the fundamental areas are covered in entirely updated or new chapters. The original text is provided as a searchable pdf file on the accompanying USB. * This book explains industrial mixers and mixing problems clearly and concisely. * Gives practical insights by the top professionals in the field, combining industrial design standards with fundamental insight. * Details applications in 14 key industries. Six of these are new since the first edition. * Provides the professional with information he/she did not receive in school. * Five completely rewritten chapters on mixing fundamentals where significant advances have happened since the first edition and seven concise update chapters which summarize critical technical information.

http://www.wiley.com/WileyCDA/WileyTitle/productCd-0470523824,subjectCd-CH30.html

http://www.amazon.com/Advances-Industrial-Mixing-Companion-Handbook/dp/0470523824/ref=sr_1_1?s=books&ie=UTF8&qid=1444143447&sr=1-1&keywords=Advances+in+Industrial+Mixing

AUTHOR INDEX

AUTHOR INDEX

AIChE
120 Wall Street, FL 23
New York, NY 10005-4020

ISBN 978-1-5108-1867-5
9 781510 818675

CURRAN ASSOCIATES INC.
proceedings.com